W0112487

SpringerBriefs in Biochemistry and Molecular Biology

For further volumes:
http://www.springer.com/series/10196

Immacolata Castellano
Antonello Merlino

Gamma-Glutamyl Transpeptidases

Structure and Function

 Springer

Immacolata Castellano
SZN
Stazione Zoologica Anton Dohrn
Naples
Italy

Antonello Merlino
Department of Chemical Sciences
University of Naples
Naples
Italy

ISSN 2211-9353   ISSN 2211-9361   (electronic)
ISBN 978-3-0348-0681-7   ISBN 978-3-0348-0682-4   (eBook)
DOI 10.1007/978-3-0348-0682-4
Springer Basel Heidelberg New York Dordrecht London

Library of Congress Control Number: 2013940951

Printed on acid-free paper

Springer is part of Springer Science+Business Media (www.springer.com)

Preface

The idea to write this book arose from the interest and discussion around our recent review covering the structural and functional aspects on gamma-glutamyl transpeptidases [1]. In the summer of 2012, Dr. Beatrice Menz, Life Sciences Editor of Springer, generously invited us to consider expanding this interesting subject and to publish it within Springer Briefs Series. We kindly accepted this invitation as we consider the topic to be of great interest for the scientific community and for a large part of readers and researchers who approach the extended literature on these enzymes.

The interest in gamma-glutamyl transpeptidases has been increased in the years due to their functional versatility and their potential applications in biotechnology and clinics. The enzymes play a key role in crucial cellular events such as aging, senescence and cancer, thanks to its involvement in glutathione metabolism, the most abundant antioxidant molecule in the cells which regulates redox homeostasis. A fascinating aspect emerging from our studies is that gamma-glutamyl transpeptidases can be considered from different point of views, a moonlighting protein, that is a protein, usually an enzyme, which can perform more than one function in different contexts, cell types, tissues, or during evolution. A number of the currently known moonlighting proteins are evolutionarily derived from highly conserved enzymes, also called ancient enzymes. Indeed, gamma-glutamyl transpeptidase is present in many different organisms and this increases the chance to develop secondary moonlighting functions. Our previous investigations on gamma-glutamyl transpeptidases isolated from extremophilic microorganisms, considered the first bacteria to colonize the earth, thanks to their ability to adapt to the extreme conditions of temperature and pressure, strongly suggest that these enzymes first evolved the ability to hydrolyze glutathione and only later with eubacteria and eukarya evolved a second function, to transfer the gamma-glutamyl group to amino acids and short peptides [2, 3]. A real gain of function that provided the cell for the uptake and recycle of amino acids for protein synthesis. An interesting prospective for this function is that the bacterial enzymes, which can accept a large spectrum of substrates (most of them still unknown), can be exploited to produce gamma-glutamyl compounds extremely useful for drug delivery. In addition, moonlighting provides a possible mechanism by which bacteria may become resistant to antibiotics. Interestingly, some mutations in bacterial gamma-glutamyl transpeptidase confer the protein cephalosporin acylase activity, a sort of antibiotic resistance, which can be exploited for an eco-sustainable synthesis of antibiotics. As regards eukaryal enzymes, gamma-glutamyl transpeptidase is also involved in leukotrienes metabolism and in drugs detoxification. All these features together make the enzyme the center of an interesting debate about its known and unknown functions.

The present volume especially focuses on the structure and function relationship of gamma-glutamyl transpeptidases and on fallout on cell life, with a particular emphasis on biotechnological and biomedical applications. In the final part, we summarize some interesting findings about the involvement of the enzyme in cancer and other diseases. For this topic, we take a lot of advantage from the more recent reviews of Pompella et al., which have given a strong contribution in understanding the role of gamma-glutamyl transpeptidase in cell proliferation. For these reasons, we thank all the researchers who have contributed with their work to our knowledge in the field.

Naples, Italy   Immacolata Castellano
                                                                                        Antonello Merlino

References

1. Castellano I, Merlino A (2012) Gamma-glutamyl transpeptidases: sequence, structure, biochemical properties, and biotechnological applications. Cell Mol Life Sci 69 (20):3381–3394
2. Castellano I, Merlino A, Rossi M, La Cara F (2010) Biochemical and structural properties of gamma-glutamyl transpeptidase from *Geobacillus thermodenitrificans*: an enzyme specialized in hydrolase activity. Biochimie 92 (5):464–474
3. Castellano I, Di Salle A, Merlino A, Rossi M, La Cara F (2011) Gene cloning and protein expression of gamma-glutamyl transpeptidases from *Thermus thermophilus* and *Deinococcus radiodurans*: comparison of molecular and structural properties with mesophilic counterparts. Extremophiles 15 (2):259–270

Contents

Gamma-Glutamyl Transpeptidases: Structure and Function

Abstract Gamma-glutamyl transpeptidases (γ-GTs) belong to the N-terminal nucleophile hydrolase superfamily, enzymes that cleave the γ-glutamyl amide bond of glutathione to give cysteinylglycine. The released γ-glutamyl group can be transferred to water (hydrolysis) or to amino acids or short peptides (transpeptidation). γ-GT plays a key role in the gamma-glutamyl cycle by regulating the cellular levels of the antioxidant molecule glutathione, hence it is a critical enzyme in maintaining cellular redox homeostasis. γ-GT is upregulated during inflammation and in several human tumors, and it is involved in many physiological disorders related to oxidative stress, such as Parkinson's disease and diabetes. Furthermore, this enzyme is used as a marker of liver disease and cancer. This book covers the current knowledge about the structure–function relationship of γ-GTs and gives information about γ-GT applications in different fields ranging from clinical biochemistry to biotechnology and biomedicine.

Keywords Gamma-glutamyl transpeptidase · Glutathione · Autoproteolytic activation · Gamma-glutamyl hydrolase · Biotechnological applications

1 Introduction

Glutathione (GSH, γ-L-glutamyl-L-cysteinylglycine) is a tri-peptide present in all mammalian tissues at 1–10 mM concentrations. It is the most abundant antioxidant molecule in cells and is involved in several crucial cellular functions, such as redox signaling, detoxification of xenobiotics and/or their metabolites, modulation of cell proliferation, apoptosis, and fibrogenesis. GSH is also a key determinant of nitric oxide (NO) and cysteine storage and transport, of sulfur assimilation, protection of cells against oxidative stress [1]. GSH cannot be hydrolyzed by general peptidases, because it contains an unusual peptide bond linking glutamate and cysteine through the γ-carboxyl group of glutamate rather than the conventional α-carboxyl group (Fig. 1).

I. Castellano and A. Merlino, *Gamma-Glutamyl Transpeptidases*,
SpringerBriefs in Biochemistry and Molecular Biology,
DOI: 10.1007/978-3-0348-0682-4_1, © The Author(s) 2013

Fig. 1 Structure of
glutathione. The amino-
terminal group of cysteine is
linked to the γ-carboxyl
group of glutamate

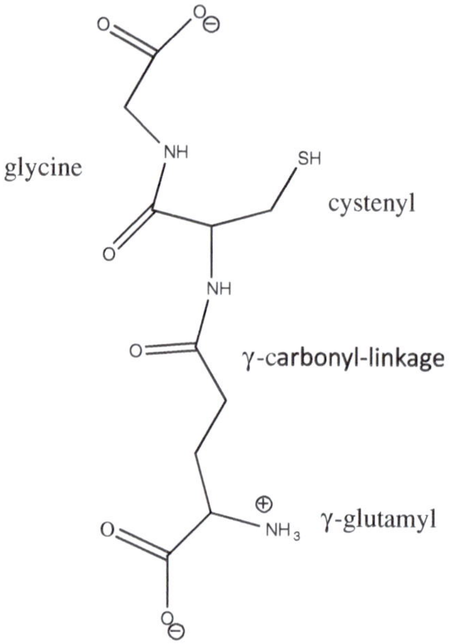

glycine

cystenyl

γ-carbonyl-linkage

γ-glutamyl

In mammalian, GSH biosynthesis [2] occurs through two ATP-dependent reactions, usually involving distinct enzymes such as glutamate-cysteine ligase, also known as γ-glutamylcysteine synthetase and glutathione synthetase (Fig. 2). The first step is catalyzed by γ-glutamylcysteine synthetase, whose structure is composed of heavy (or catalytic) and small (modifier) subunits, which are encoded by different genes in humans. This step conjugates cysteine with glutamate and generates γ-glutamylcysteine. In the second step, glutathione synthetase (GSH synthetase) leads to the production of GSH from γ-glutamylcysteine and glycine. In humans, GSH synthetase is a homodimer.

The γ-glutamyl cycle utilizes GSH as a continuous source of cysteine for the cell (Fig. 3). It is synthesized in the cytosol and is then translocated out of the cell. Extracellular GSH metabolism is initiated by the γ-Glutamyl Transpeptidase (γ-GT, EC 2.3.2.2), the first enzyme of the GSH degradation pathway, and then completed by membrane dipetidases (Fig. 3).

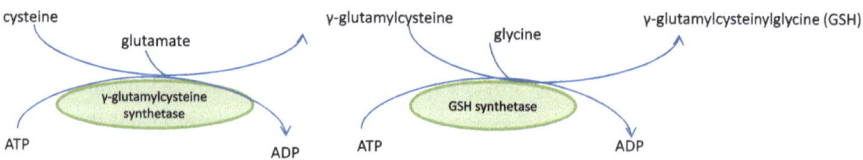

Fig. 2 GSH synthesis. GSH synthesis occurs via a two-step ATP-requiring enzymatic process. The first step is catalyzed by γ-glutamylcysteine synthetase. The second step is catalyzed by GSH synthetase. GSH exerts a negative feedback inhibition on γ-glutamylcysteine synthetase

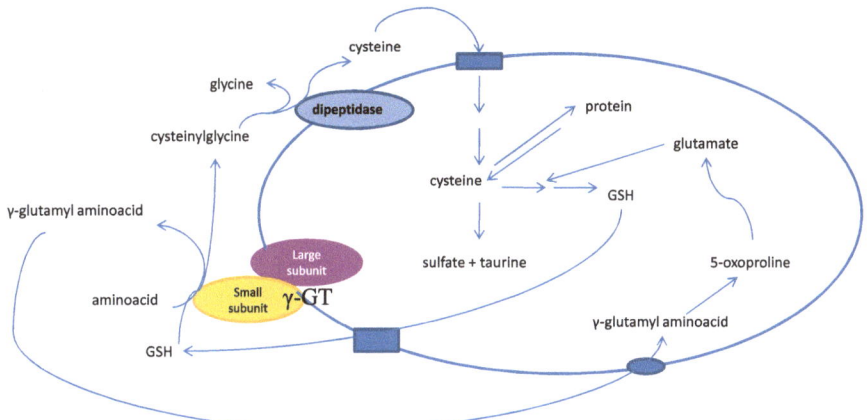

Fig. 3 γ-glutamyl cycle. GSH is transported out of the cell where γ-GT, which is constituted by a large and a small subunit, transfers the γ-glutamyl moiety of GSH to an amino acid (acceptor), forming γ-glutamyl amino acid and cysteinylglycine. The γ-glutamyl amino acid can then be transported back into the cell. Once inside, it can be further metabolized to release the amino acid and 5-oxoproline, which can be converted to glutamate and reincorporated into GSH. Cysteinylglycine is broken down by dipeptidases to generate glycine and cysteine, which is also transported back into the cell. Most of the cysteine taken up by the cell is reincorporated into GSH while the remaining amount is incorporated into newly synthesized proteins and/or broken down into sulfate and taurine. Figure adapted from Lu [192]

γ-GT (E.C.2.2.3.2) is an evolutionary conserved enzyme that specifically cata-lyzes the cleavage of the γ-glutamyl bond of GSH and the transfer of the γ-glutamyl group to water (hydrolysis), amino acids, or peptides (transpeptidation) [3, 4].

Mammalian γ-GT is a glycoprotein integrated in the plasma membrane with its active site facing the extracellular space, where γ-glutamyl moieties of GSH are likely hydrolyzed and transferred to other amino acids, leading to the formation of γ-glutamyl amino acids, which are then transported into the cell. The cleavage of the γ-glutamyl bond of extracellular GSH enables the cell to use this antioxidant compound as a source of cysteine, which in turn serves as essential substrate for intracellular glutathione and protein synthesis. Under this concern, γ-GT-mediated metabolism of extracellular GSH is especially critical for protein synthesis in rapidly dividing neoplastic cells [5].

In this book, we try to connect structural and functional data on γ-GTs from different sources, with the aim to stimulate interest and provide an overview for future research on γ-GTs.

2 Occurrence, Genes, and Gene Expression

High levels of γ-GT are constitutively expressed in adult animals, and enzymatic activity is relatively higher in the kidney, intestine, and epididymis [6]. In partic-ular, the enzyme is the most abundant in the proximal tubule of the kidney. The expression of γ-GT is regulated in a tissue-specific manner and is closely related to differences in glutathione metabolism. In general, the enzyme is often expressed in tissues that are involved in the transport of different biological compounds, and the activity of the enzyme is relatively high in the luminal surface of ductal tissues. In γ-GT-deficient mice [7], the lack of γ-GT leads to the excretion of large amounts of cysteine in the urine, thus suggesting that the kidney γ-GT plays a critical role in the recovery of cysteine from urinary glutathione. As described above, the availability of cysteine for protein synthesis and other metabolic pathways is greatly dependent on cellular GSH levels rather than on free cysteine and cystine, thus, glutathione represents a nutrient source for the cell. Indeed, the reabsorption of the cysteine from urine occurs by the action of kidney γ-GT. Most cells are not able to directly uptake both reduced and oxidized GSH, thus γ-GT makes cysteine, contained in extracellular GSH, available to cells. For this reason, γ-GT seems to be associated with drug resistance in some cancer cells, in which the enzyme is up-regulated [8].

GSH is synthesized by almost all mammalian cells, but the liver is the tissue which produces the largest amount. This tissue is relatively unique in terms of the absence of γ-GT hydrolase activity toward glutathione. In contrast to tissues such as the kidney and intestine, the activity of γ-GT is low and nearly undetectable in the adult liver. However, a transient increase in γ-GT activity is observed in the liver at the perinatal stage during development [9]. On the other hand, the activity of the enzyme in the kidney is initially low in the neonate, then the enzyme level increases during the neonate growth and reaches a maximal level at maturity. In both rats and mice, although fetal liver hepatocytes express the enzyme at significant levels, γ-GT expression is suppressed after birth. Together with gene suppression, the activity becomes lower and finally decreases to undetectable levels. The different profiles of γ-GT activity during embryonic development vary among different species. It has been proposed that the glucocorticoid hormone plays a role in the occurrence of this increase [10]. Because this transient increase in γ-GT activity occurs in concomitant with the active growth of cells in the fetal liver, it was proposed that two events are associated [11].

2.1 Hepatic Expression of γ-GT

An alteration in the activity of γ-GT in the liver is observed during development from the embryo to the neonate and in chemical carcinogenesis. In the development of the liver, γ-GT activity is transiently increased at the perinatal stage, whereas it is restricted to biliary cells in the adult liver. Thus, from fetal development to the adult stage, γ-GT is expressed essentially in biliary cells and immature perinatal hepatocytes [9]. The expression of γ-GT in liver and other organs such as the kidney is distinctly regulated. Although in the kidney a large amount of mRNA is transcribed by the most proximal promoter, in biliary cells and in immature hepatocytes the gene is transcribed by different promoters [12]. During liver development, immature hepatic cells differentiate into hepatocytic and biliary lineages. These bipotential precursor cells alter the gene expression pattern of γ-GT and the use of promoters during differentiation.

2.2 γ-GT Gene Structure

γ-GTs are encoded by a single gene and are translated as a unique polypeptide chain, which then undergoes an autoproteolytic cleavage into a heavy and a light chain, usually called large and small subunit, respectively (see below). The γ-GT gene was found as a single copy in both rat and mouse, whereas the human counterpart was found as a multicopy gene. The rodent γ-GT genes contain 12 coding exons, interrupted by 11 introns. The organization of the γ-GT gene is well conserved in rat and mouse, and multiforms of mRNA arise from the single copy gene in a tissue-specific manner. Multipromoter and splicing variations allow transcriptions of several species of mRNA for γ-GT [13]. In the rat, five different promoters have been identified, and, at least six transcripts, including a splicing variant, have been characterized. The tissue-specific expression of γ-GT is conferred by multiple promoters, even though the different transcripts share an identical coding sequence. Promoters-I, -II, and -IV are active in the rat kidney, small intestine, and epididymis, where the highest enzyme activity is found, and seem to serve for the constitutive expression of γ-GT. The promoter-II contains a TATA-like sequence at the position expected from the transcription startsite. These findings have revealed that the tissue-specific expression of γ-GT is conferred by multiple promoters, even though all the different transcripts share an identical coding sequence.

The human multigene family coding for γ-GT encompasses at least seven different genes [14]. Most members of the gene family are localized on chromosome 22 [15]. Human γ-GT (hGT) is also transcribed under the control of multipromoters, and several mRNA species have been identified. Alternative splicing and transcription from multiple genes also contribute to the diversity of the transcripts [16]. Some transcripts do not code for an entire protein sequence but for truncated forms of the protein, in which the open reading frame comprises sequences that largely consist of either large or small subunits. Translation has been found to be regulated by the $5'$ untranslated region of an mRNA for γ-GT from a human hepatocellular carcinoma cell line, HepG2 cells, and thus, the region seems to serve as a tissue-specific active translational enhancer [17]. Such a regulating mechanism for translational efficiency together with transcriptional regulation by the multipromoter system would confer more flexible control of the expression of γ-GT.

2.3 Induction of Hepatic γ-GT

An increase in mRNA levels and enzyme activity has been found in the liver of adult animals in response to the administration or intake of xenobiotics, carcinogens, and alcohol (see also γ-GT in bioclinical). Drugs cause two peaks of

γ-GT activity. The former is reversible as γ-GT returns to basal levels after stopping the administration. Instead, the latter is irreversible and independent from the discontinuation of the chemical. This γ-GT profile seems to be consistent with the Solt-Farber's initiation/promotion model of chemical carcinogenesis [18, 19]. The multistep process of liver carcinogenesis encompasses three different steps: initiation, promotion and progression, and the proceeding reversible elevation of γ-GT activity corresponds to the initiation step. At this state, γ-GT-positive foci and hyperplastic nodules are observed in the liver, and it is thought that such histological alterations can be extinguished by redifferentiation of the cells and tissue remodeling. The late irreversible increase in enzyme activity seems to be an event associated with carcinogenesis. Besides a number of chemical carcinogens, other chemical compounds are also able to induce the hepatic expression of γ-GT. For example, phenobarbital, polychlorobiphenyls, and a choline-deficient/low-methionine diet promote the initiation/promotion model in the liver [18, 19].

2.4 γ-GT-Related Gene Products

On the basis of primary structure similarity, an enzyme structurally related to γ-GT has been identified in humans. This enzyme, named as glutamyl-leukotrienase, catalyzes the hydrolysis of glutathione and a glutathione-conjugate, leukotriene. However, γ-glutamyl-p-nitroanilide, the synthetic substrate widely used in the assay of γ-GTs (see enzymatic assays), does not serve as a substrate for this enzyme. Glutamyl-leukotrienase from mouse seems to be very similar to its human counterpart. However, the enzymatic activity of these γ-GT-like proteins are different in terms of substrate specificity (e.g., the mouse enzyme is not capable of hydrolyzing GSH). Glutamyl-leukotrienase is abundant in the spleen and is involved in the inflammatory response. The discovery of structurally and functionally related γ-GT-enzymes has led to the suggestion that γ-GT is not only involved in glutathione metabolism but that it can play other roles to finely respond to different biological processes. Indeed, it has been reported that γ-GT, despite its enzymatic activity toward GSH, can induce osteoclast formation in bone marrow cultures [20].

3 Structure and Function of γ-GTs

The first γ-GT was isolated from sheep kidney in 1950 [21]. Today the primary structures of γ-GTs from various sources, ranging from bacteria to mammals, are known. As already mentioned, γ-GT polypeptide chain undergoes an autoproteolytic cleavage into a large and a small subunit. Sequence alignments point to a strong conservation of structure and function. Indeed, these enzymes share >25 % identity, with the sequence of the small subunit slightly more conserved than that of the large subunit (Figs. 4 and 5) [22]. From a multiple alignment of 501 γ-GT sequences, about 80 residues appear highly or partially conserved. Part of these residues is responsible for the correct protein folding, others are essential for substrate recognition and needed to maintain the hydrolase activity, and other residues are conserved for unknown aspects. Among the conserved positions, some important residues have been identified: (i) a Thr (T391 in *Escherichia coli* γ-GT, EcGT; T381 in human GT, hGT), the nucleophile responsible for both the auto-proteolysis and the enzymatic activity and (ii) two Gly residues (G483-G484 in EcGT, G473-G474 in hGT) proposed to have a role in the stabilization of the tetrahedral transition state of the enzyme (Fig. 4) [22, 23].

The molecular weights of the two chains are generally found to be within 38–72 kDa for the large subunit and within 20–66 kDa for the small subunit [24–27]. These variations in the molecular masses are due to the high glycosylation levels of the animal and plant enzymes [28–30]. Glycosylation is believed to confer protection against proteases. In mammalian, the large subunit has an intracellular N-terminal sequence, a single transmembrane domain anchored to membrane by a hydrophobic fragment of eight residues and a large extracellular domain that binds the small subunit. The extracellular component can be separated from the membrane by papain treatment [31]. The hydrophobic anchor can be removed without affecting the enzyme activity [32]. Bacterial homologs occur as soluble proteins, localized in the periplasmic space [25] by an N-terminal signal peptide or secreted in the extracellular space [33].

hGT has a variety of isoforms that differ in their carbohydrate content and structure. It contains seven N-glycosylation sites [34]: six sites in the large subunit (N95, 120, 230, 266, 297, and 344) and one in the small subunit (N511). Four N-glycosylation sites (N95, 120, 344, and 511) are highly conserved among eukaryotes, two additional ones (N230 and 297) are conserved among mammalian γ-GTs. The remaining *N*-glycosylation site (N266 in hGT) is not conserved: for example, it is not present in pig and rat. C50 and C74, and C192 and C196 most likely form disulphide bonds. The C192–C196 pair is only found in a subset of mammalian γ-GTs [35], whereas the C50–C74 pair is highly conserved in non-bacterial homologs. Since these residues are not involved in the catalytic mechanism, they could have a structural role and/or a regulatory function [23].

X-ray crystallography has played a crucial role in the study of γ-GTs, providing researchers with detailed structural models that have been essential to unveil the enzyme function and reactivity. Until now, structural investigations of γ-GTs have been mostly concentrated on bacterial enzymes, with just 15 structures deposited in the Protein Data Bank (PDB, Table 1), probably because of difficulty in crystallization of the heavily and heterogeneously glycosylated mammalian proteins. Okada et al. [22] published the first crystal structure of a γ-GT in 1996, that is, the structure of EcGT (PDB code 2DBU). EcGT crystals were grown from 15 % PEG 4,000, 0.2 M magnesium sulfate, 5 % glycerol, at pH 5.0 and at a temperature of 277 K. They were shown to diffract to a resolution of 1.95 Å. Since then, the homologous proteins from *Helicobacter pylori*, HpGT (PDB code 2NQO) [23], and from *Bacillus subtilis* GT (BsGT) (PDB code 2V36) have also been

```
hGT         TAHLSVVAEDGSAVSATSTINLYFGSKVRSPVSGILFNNEMDDFSS-PSITNEFGVPPSP 439
ratGT       TAHLSVVSEDGSAVAATSTINLYFGSKVLSRVSGILFNDEMDDFSS-PNFTNQFGVAPSP 438
D.rerioGT   TAHLSVIAEDGSAVAATSTINLYFGSKVMSRSTGIIFNDEMDDFSS-PYITNGFGVPPSP 359
D.melaGT    TAHMNVLATNGDAVSITSTINNYFGSKVASTQTGIILNDEMDDFST-PGVINGFGVPASP 440
S.cerGT     TAHFSIVDSHGNAVSLTTTINLLFGSLVHDPKTGVIFNNEMDDFAQ-FNKSNSFELAPSI 528
EcGT        TTHYSVVDKDGNAVAVTYTLNTTFGTGIVAGESGILLNNQMDDFSAKPGVPNVYGLVGGD 450
HpGT        TTHYSVADRWGNAVSVTYTINASYGSAASIDGAGFLLNNEMDDFSIKPGNPNLYGLVGGD 439
BsGT        TTHFTVADRWGNVVSYTTTIEQLFGTGIMVPDYGVILNNELTDFDA---------IPGG 452
BlGT        TTHFTVTDQWGNVVSYTTTIEQLFGTGILVPGYGLFLNNELTDFDA---------IPGG 448
GthG        TTVYLAAADGEGNMVSFIQSNYMGFGSGLVVPGTGIALHNRGHNFVF---------DENH 402
DrGT        TVVYLAAADDEGGMVSMIQSNYMGFGSGVVVPGTGIALHNRGHNFHT---------DPAH 402
TtGT        TVVYLAAADGE-VMVSLIQSNYQGFGSGVLVPGTGIALQNRGLGFSL---------EEGH 394
TaGT        TTYFSISDSEGRSVSIIQSNYMGFGSGIVPKGTGFVLQNRGSYFTL---------QRDH 380
              *                                                   _____
```

```
hGT         ANFIQPGKQPLSSMCPTIMVGQDGQVRMVVGAAGGTQITTATALAIIYNLWFGYDVKRAV 499
ratGT       ANFIKPGKQPLSSMCPSIIVDKDGKVRMVVGASGGTQITTSVALAIINSLWFGYDVKRAV 498
D.rerioGT   NNFIQPGKRPLSSMCPTIIFDKHNRVKMVVGASGGTKITTATALVILNSLFFNYDLKKAV 419
D.melaGT    ANYIYPGKRPMSSMSPCIIVDQEGNVRLLVGAAGGTRIITSVAAVIMKYLLRKESLTAAV 500
S.cerGT     YNFPEPGKRPLSSTAPTIVLSELGIPDLVVGASGGSRITTSVLQTIVRTYWYNMPILETI 588
EcGT        ANAVGPNKRPLSSMSPTIVVKD-GKTWLVTGSPGGSRIITTVLQMVVNSIDYGMNVAEAT 509
HpGT        ANAIEANKRPLSSMSPTIVLKN-NKVFLVVGSPGGSRIITTVLQVISNVIDYNMNISEAV 498
BsGT        ANEVQPNKRPLSSMTPILFKD-DKPVLTVGSPGGATIISSVLQTILYHIEYGMELKAAV 511
BlGT        ANEVQPNKRPLSSMTPTIVFKD-EKPVLTVGSPGGTTIIASVFQTILNYFEYGMSLQDAI 507
GthGT       PNGLAPRKKPYHTIIPGFLTKGG-KPIGPFGVMGGFMQPQGHMQVIMNTVDFALNPQAAL 461
DrGT        PNALAPGKRPYHTIIPGFLGRADGTPVGPFGVMGGFMQPQGQLQVVVNTVRYGMNPQQAL 462
TtGT        PNRVGPGKRPFHTIIPGFLAREG-KPLGPFGVMGGFMQPQGHVQVVVGLADFGLNPQAAL 453
TaGT        PNALMPGKRTFHTLAACMVEKEH-DLYASLGSMGGDIQPQVQMQILMEILKDNTDPQAIL 439
                                                             **
```

```
hGT         EEPRLHNQLLPNVTTVERN--IDQAVTAALET------RHHHT---QIASTFIAVVQAIV 548
ratGT       EEPRLHNQLLPNTTTVEKN--IDQVVTAGLKT------RHHHT---EVTPDFIAVVQAVV 547
D.rerioGT   TEPRVHNQLNPNMTVVEQD--FEQSVLDGLEQ------KNHVT---ELQRTPGAVVQAIV 468
D.melaGT    NNGRLHHQLAPMRVSYEQE--VDSSVTDYLKQ------VGHEMYE-EPVGSSFAAVTAIG 551
S.cerGT     AYPRIHHQLLPDRIELESFPMIGKAVLSTLKE------MGYTMK--EVFPK--SVVNAIR 638
EcGT        NAPRFHHQWLPDELRVEK--GFSPDTLKLLEA------KGQKVALKEAMGSTQSIMVGPD 561
HpGT        SAPRFHMQWLPDELRIEKF-GMPADVKDNLTK------MGYQIVTKPVMGDVNAIQVLPK 551
BsGT        EEPRIYTNSMSSYRYEDG---VPKDVLSKLNG------MGHKFGT-SPVDIGNVQSISID 561
BlGT        EEPRIYTNSLTSYRYESG---MPEDVRRKLND------FGHKFGS-NPVDIGNVQSIFID 557
GthGT       DAPRWQWMEGKTVLVEPH---FPRHIAEALAR------KGHDICVALDGGPFGRGQIIWR 512
DrGT        DAPRWQWLQGRTVEVEPA---LGDQLARALVA------RGHDVRVQLDPGSFGRGQMIRR 513
TtGT        DRPRWQVVPGDEVLLEPG---IPQATALFLKD------LGHRVRYEAEYGLFGRGQVVFR 504
TaGT        DKPRWTEP--YTIYEAPG---AVYVESEELYRNVSKQISGRKVVLRDVSQEFGTAQITTL 494
```

```
hGT         RTAGGWAAASD-SRKGGEPAGY-------- 569
ratGT       RTSGGWAAASD-SRKGGEPAG--------- 568
D.rerioGT   RQGDKLCAECD-PRKGGYPAGY-------- 489
D.melaGT    -ALEQPEPFYD-RRRIGSALTLAKTNKMQH 579
S.cerGT     NVRGEWHAVSDYWRKRGISSVY-------- 660
EcGT        GE---LYGASDPRSVDDLTAGY-------- 580
HpGT        TKGSVFYGSTDPR--KEF----------- 567
BsGT        HENGTFKGVADSSRNGAAIGINLKRK---- 587
BlGT        RENKTFMGVADSSGNGTAVGVNNKTSAE-  585
GthGT       DPDTGVLAAGTEPRTDGAVAAW-------- 534
DrGT        DPDTGVLEGGTESRTDGHIALW-------- 535
TtGT        --LGEALVGASDPRAEGLALAW-------- 524
TaGT        -IRGDVVVGAADPRGDGIAIPYS------- 516
```

structurally characterized. The structures of the putative γ-GT from *Bacillus halodurans* (BhGT, PDB code 2NLZ, annotated as cephalosporin acylase) and from the thermoacidophilic archaeon *Thermoplasma acidophilum* (TaGT, PDB code 2I3O) have been deposited in the PDB without a primary citation. Furthermore, it has been recently solved the structure of CapD from *Bacillus anthracis*, a protein considered related to γ-GTs, but not able to hydrolyze GSH (PDB code 3G9K) [36]. Crystallization conditions of these proteins are reported in Table 2.

◀ **Fig. 4** Multiple alignment of amino acid sequences of the small subunits of some γ-GTs. The amino acid sequences of γ-GT from *H. sapiens* (Swiss-Prot P19440), *R. norvegicus* (Swiss-Prot P07314), *D. rerio* (Swiss-Prot Q7T2A1), *D. melanogaster* (Swiss-Prot Q9VWT3), *S. cerevisiae* (Swiss-Prot Q05902), *E. coli* (Swiss-Prot P18956), *H. pylori* (Swiss-Prot O25743), *B. subtilis* (Swiss-Prot P54422), *B. licheniformis* (Swiss-Prot Q62WE3), *G. thermodenitrificans* (YP001127364.1), *D. radiodurans* (Gene ID: 1798291), *T. thermophylum* (Gene ID: 3168743) and *T. acidophilum* (NP394454.1) are included. The alignment was performed using the ClustalW method (http://www.ebi.ac.uk/Tools/msa/clustalw2/). The conserved threonine responsible for autoprocessing of the enzyme and the two strictly conserved glycines involved in binding of the γ-glutamyl moiety (T391, G483, and G484 in EcGT) are indicated by an asterisk. The lid loop extending toward the active site (spanning from P438 to G449 of EcGT) and absent in some γ-GTs is underlined

All these γ-GTs exhibit a similar structure: they belong to the superfamily of the N-terminal nucleophilic (Ntn) hydrolase [37, 38]. The members of this superfamily share a common active site architecture, tertiary and quaternary fold, although they can show pairwise sequence identities as low as 15 %. The predominant elements of secondary structure in Ntn-hydrolases are a long four-layer αββα-structure, with two antiparallel β-sheets packed against each other and sandwiched between α-helical layers (Fig. 6a, b). In the central β-sheet sandwich, both large and small subunits provide strands. They are arranged in an antiparallel fashion: one of the two central sheets is essentially flat, whereas the other may be twisted at the end (Fig. 6b).

In the studied γ-GTs, the small and the large subunits are highly intertwined throughout the structure. The inter-subunit interface is stabilized by both hydrogen bonds and hydrophobic contacts. The crystal structure of EcGT has been solved

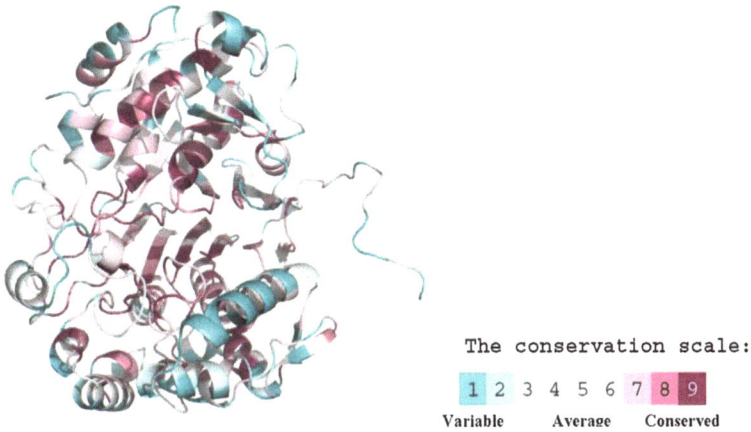

The conservation scale:

1 2 3 4 5 6 7 8 9

Variable   Average   Conserved

Fig. 5 Ribbon representation of one subunit of the precursor structure of the γ-GT from *B. halodurans* (PDB code 2NLZ) colored by residue conservation. Residues were colored according to their evolutionary conservation score using the color-coding bar on the right, as implemented in the *CONSURF* web server (http://consurf.tau.ac.il/)

Table 1 Crystal symmetry and unit cell dimensions for γ-GT structures

Enzyme	Structure description	PDB code	Resolution (Å)	Space group	Unit cell dimension a,b,c (Å) α,β,γ (°)
*Ec*GT	Ligand free form	2DBU	1.95	P2$_1$2$_1$2$_1$	78.7, 126.9, 128.8 90, 90, 90
*Ec*GT	Complex with hydrolyzed GSH	2DG5	1.60	P2$_1$2$_1$2$_1$	78.1,126.6, 129.2 90, 90, 90
*Ec*GT	Complex with Acivicin	2Z8K	1.65	P2$_1$2$_1$2$_1$	77.7,126.5, 129.4 90, 90, 90
*Ec*GT	Complexed with L-Glu	2DBX	1.70	P2$_1$2$_1$2$_1$	77.7,126.5, 129.2 90, 90, 90
*Ec*GT	Complex with Azaserine	2Z8I	1.65	P2$_1$2$_1$2$_1$	77.5,126.2, 129.2 90, 90, 90
*Ec*GT	Acyl-Enzyme Intermediate	2DBW	1.80	P2$_1$2$_1$2$_1$	78.8,126.7, 128.9 90, 90, 90
*Ec*GT	Monoclinic form	2E0X	1.95	P2$_1$	65.2,127.9, 75.1 90, 94.8, 90
*Ec*GT	Sm derivative	2E0Y	2.02	P2$_1$2$_1$2$_1$	78.9,125.9, 128.9 90, 90, 90
*Ec*GT	Complex with Azaserine in the dark	2Z8J	2.05	P2$_1$2$_1$2$_1$	76.3,126.8, 128.1 90, 90, 90
*Ec*GT	T391A mutant	2E0W	2.55	P4$_3$2$_1$2	134.6,134.6, 118.4 90, 90, 90
*Hp*GT	Ligand free form	2NQO	1.90	P2$_1$	54.3,105.2, 91.1 90, 92.0, 90
*Hp*GT	Complex with Glu	2QM6	1.60	P2$_1$	55.0,104.8, 91.9 90, 91.8, 90
*Hp*GT	Complex with Acivicin	3FNM	1.70	P2$_1$	54.7,105.4, 91.9 90, 91.7, 90
*Hp*GT	Mature T380A with S(nitrobenzyl)GSH	2QMC	1.55	P2$_1$	57.1,106.7, 87.2 90, 105.0, 90
*Bs*GT	Ligand free form	2V36	1.85	P2$_1$2$_1$2$_1$	72.3,108.8, 161.3 90, 90, 90
*Bs*GT	Complex with glutamate	3A75	1.95	P2$_1$2$_1$2$_1$	49.4, 98.9, 227.2 90, 90, 90
*Ta*GT	Ligand free form	2I3O	2.03	P2$_1$	116.1,95.5, 119.0 90, 109.8, 90
*Bh*GT	Ligand free form	2NLZ	2.70	P3$_1$2	105.7,105.7, 385.1 90, 90, 120
CapD *B. anthracis*		3G9K	1.99	P2$_1$	52.7,120.6, 77.4 90, 90.9, 90

Table 2 Crystallization conditions of γ-GTs

Protein	Technique	pH	Buffer/Salt	Precipitant	Additive/Cryoprotectant	Temperature (K)
EcGT	Vapor diffusion Hanging drop	5	0.2 M Mg sulfate,	15 % PEG 4,000	5 % glycerol	277
		8.5	0.2 M CaCl₂, 0.1 M Tris-HCl	20 % PEG 4,000		
HpGT	Vapor diffusion Sitting drop	7.5	200 mM HEPES	25 % MPEG2000		291
BsGT	Vapor diffusion Hanging drop	7	100 mM MES	20 % PEG 4,000, 600 mM NaCl	5 % Jeffamine M-600	293
TaGT	Vapor diffusion Hanging drop	5.5	Mg Formate			293
BhGT	Vapor diffusion Sitting drop	7.5	200 mM Na Formate	20 % PEG 3,350	20 % glycerol	294

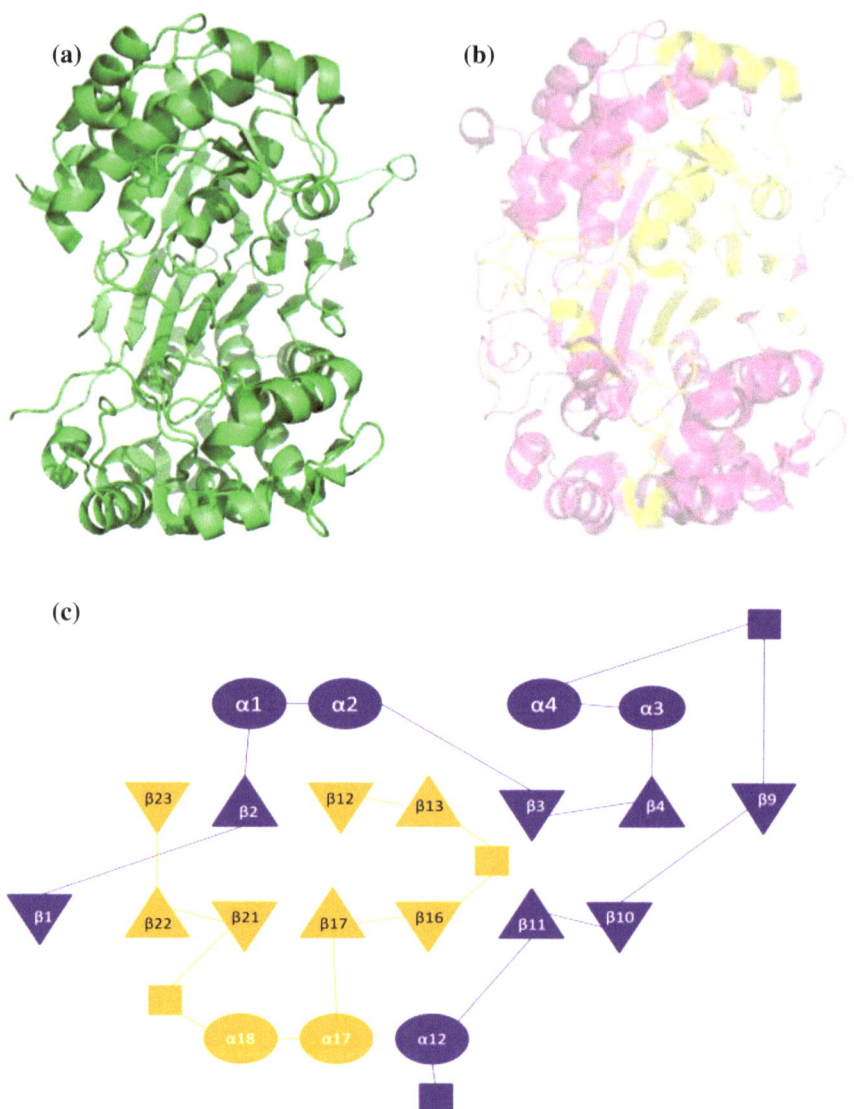

Fig. 6 Ribbon representations of the structure of (**a**) precursor (*green*) and (**b**) mature form of EcGT. In the latter, small and large subunits are shown in *yellow* and *purple*, respectively. (**c**) A topology diagram of EcGT showing the disposition of secondary structure elements in the protein. In this diagram, *circle*, *triangle*, and *square* indicate α-helix, β-strand and not conserved insertions, respectively

both for the unprocessed T391A mutant (PDB code 2E0W) [39] and for the mature enzyme (PDB code 2DBU) [22]. A detailed comparison of these structures is reported below.

An important contribution to the understanding of γ-GT function has been the elucidation of X-ray structures of EcGT and BsGT complexes with L-glutamate, which acts as substrate/product analog (PDB codes 2DBX and 3A75, respectively [22, 40]). Structures of γ-GTs in complex with ligands include those of HpGT with Acivicin (PBD code 3FNM) [41] and of EcGT with Azaserine (PDB code 2Z8I) or Acivicin (PDB code 2Z8K), two potent inhibitors of the protein (see later).

4 Autoprocessing and Reaction Mechanism of γ-GTs

4.1 Autoprocessing

Members of the Ntn-hydrolase superfamily are produced as inactive proenzymes (also called precursors or zymogens) that undergo a post-traslational modification to form catalytically competent enzymes. This process is a proteolytic activation that releases a catalytic serine, threonine, or cysteine at the N-terminal position [37, 38, 42] (Fig. 7).

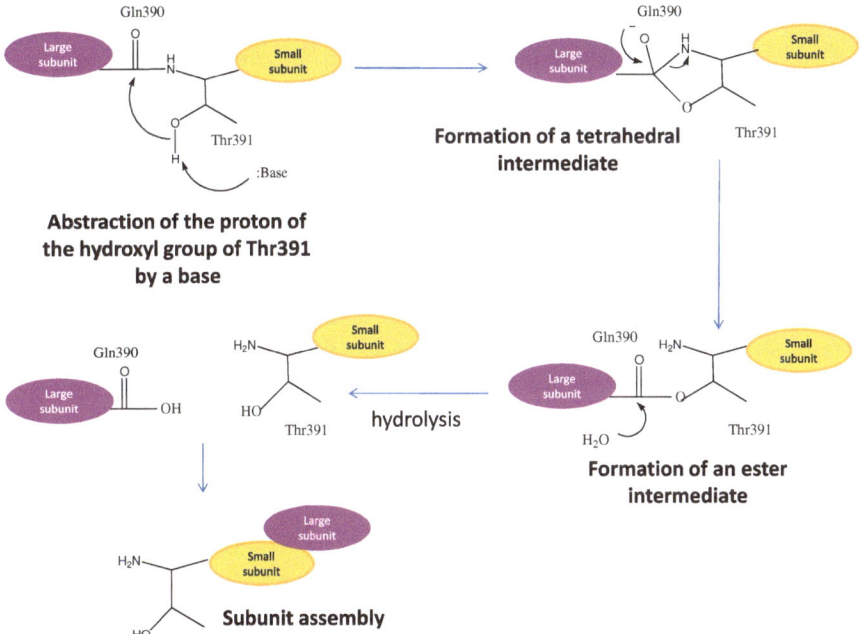

Fig. 7 Proposed mechanism for intramolecular autocatalytic processing of γ-GT from *E. coli*. A base activates Thr OG atom thereby enabling it to attack the preceding residue on C; the resultant tetrahedral transition state collapses into an ester intermediate; hydrolysis of the ester separates the two chains. Adapted from Castellano and Merlino [102]

In detail, in γ-GTs, the hydroxyl group of the strictly conserved threonine (T391 in EcGT, T381 in hGT) of the proenzyme serves as nucleophile for the cleavage. Thr attacks the carbonyl group of the preceding residue, which is often a glutamine (Gln390 in EcGT) or a glycine, forming a tetrahedral intermediate. The cleavage of the C–N bond through protonation of the amino group of the Thr yields an ester intermediate (N–O acyl shift), which is then hydrolyzed by a water molecule to form a large and a small subunit with the Thr as the new N-terminal residue. A crucial role in this process could be played by a water molecule, which could enhance the nucleophilicity of the hydroxyl group of the Thr [39, 43, 44]. In hGT, glycosylation is required to process the proenzyme, even though nothing is known about the role of glycosylation in the autoprocessing [34, 45].

The precursor and the mature (and active) γ-GTs have distinct structural features in different organisms. In mammalians, as well as in many bacteria like *E. coli*, the proenzyme is a dimer of two identical monomers $(\alpha\beta)_2$, whereas the activated γ-GT is a heterodimer of the small (α-) and large (β-) subunits (Fig. 6b) [22]. In this notation, the term $\alpha\beta$ refers to the single unprocessed chain, whereas α- and β- are the two chains after processing. In the case of the thermophilic enzyme from *Geobacillus thermodenitrificans* (GthGT), the inactive precursor exhibits a homotetrameric $(\alpha\beta)_4$ structure, whereas the mature enzyme behaves as a heterotetramer $(\alpha_{-2}\beta_{-2})$ [46]. Moreover, crystal structures of both EcGT and HpGT revealed the presence of two heterodimeric molecules, that is, a hetero-tetramer, in the asymmetric unit [22, 23]. Interestingly, gel filtration and dynamic light scattering data on HpGT [47] also indicate the existence of heterotetrameric assemblies. The effect of oligomerization of γ-GTs is unknown, and its role in catalytic function and structural stabilization needs further studies. In HpGT, EcGT, and BlGT [39, 43, 44], the substitution of the conserved Thr with an Ala results in a homodimeric protein unable to undergo the autocatalytic maturation, whereas in GthGT [46] the same replacement leads to a homotetrameric proen-zyme, which keeps a reduced but significant hydrolase activity. It has been sug-gested that this small catalytic activity might be due to a water molecule that mimics the side chain hydroxyl group of the catalytic threonine [46]. In HpGT, removal of the CH_3 group of the Thr, that is, substitution of the catalytic Thr with Ser, results in a nearly 80-fold reduction in the rate of processing when compared to the wild-type enzyme. In this case, the protein needs more than 30 days for complete maturation [23, 47]. In EcGT, when Thr391 is replaced by Ser or Cys, the derived precursors are processed, albeit slowly [48].

The role of the residue preceding the Thr in the autocatalytic processing of γ-GTs is controversial. It was investigated by site-directed mutagenesis in the case of EcGT [49, 50] and BlGT [51]. This residue, which in BlGT is a glutamate, was substituted by either A, D, R, or Q. Analysis of SDS Polyacrylamide Gel Electrophoresis showed that the E → A, E → D, and E → R substitutions yield mutants unable to undergo cleavage. However, when glutamate was replaced by a glutamine, the mutant was able to process itself into a large and a small subunit, thus acquiring a catalytic activity comparable to that of the native enzyme [51]. On the contrary, studies on EcGT showed that Gln390 has no role in autoprocessing since its substitution with A does not produce any maturational defect [49, 50]. Other controversial results were obtained when a His (His393 of EcGT) rather well conserved in γ-GTs was replaced: the H → G substitution yields to the formation of unprocessed precursors in the case of EcGT [52], whereas the corresponding hGT mutant (H383A) forms a mature protein [53]. This residue is a Y in GthGT, BsGT, TtGT and TaGT (Fig. 4).

The C-terminal region of γ-GTs has also been demonstrated to have a critical role in their autocatalytic capability [43]. In particular, site directed mutagenesis and deletion studies on BlGT and HpGT have revealed that the C-terminus participates in an undefined way to the autocatalytic process and that the deletion of nine residues significantly affects the processing. In BlGT, V576 (P566 in hGT and T577 in EcGT) is located at the boundary [43].

Structural information about the inactive precursor has also been obtained by exploring the mutagenized native polypeptides with a prevented or delayed activation site. Comparison of the overall structure of the unprocessed T391A mutant of EcGT (PDB accession code: 2E0W) with that of the mature enzyme shows no major differences between the backbone atom positions, especially in the core regions [39]. Large changes are found near the active site: in fact, in the precursor analog, the segment corresponding to the C-terminal region of the large subunit occupies the site where a loop (residues 438–449 in EcGT, hereafter denoted as lid loop) forms the lid of the γ-glutamyl group-binding pocket in the mature γ-GT (Fig. 8). This feature suggests that, upon cleavage of the N-terminal peptide bond of T391, the newly produced C-terminus of the large subunit (residues 375–390 in EcGT), hereafter denoted as P-segment, flips out, allowing the formation of the γ-glutamyl group-binding pocket. The conformational variation of the P-segment can be described as a rotation about residue Ile378, which changes its ψ angle from 45° to 127°, thus allowing the displacement of the segment from the catalytic Thr (Fig. 9a, b). The rearrangement of the mobile P-segment upon the autoprocessing is favored by the formation of electrostatic interactions. The P-segment conformational changes could not be a general property of γ-GTs, since the position of

Fig. 8 Superimposed structures of inactive precursor (*green*) and mature form of EcGT. In the mature enzyme, the large subunit is shown in *violet* and the small subunit in *yellow*. γ-glutamyl binding site, catalytic Thr, P-segment, and lid loop have been evidenced. Adapted from Castellano and Merlino [102]

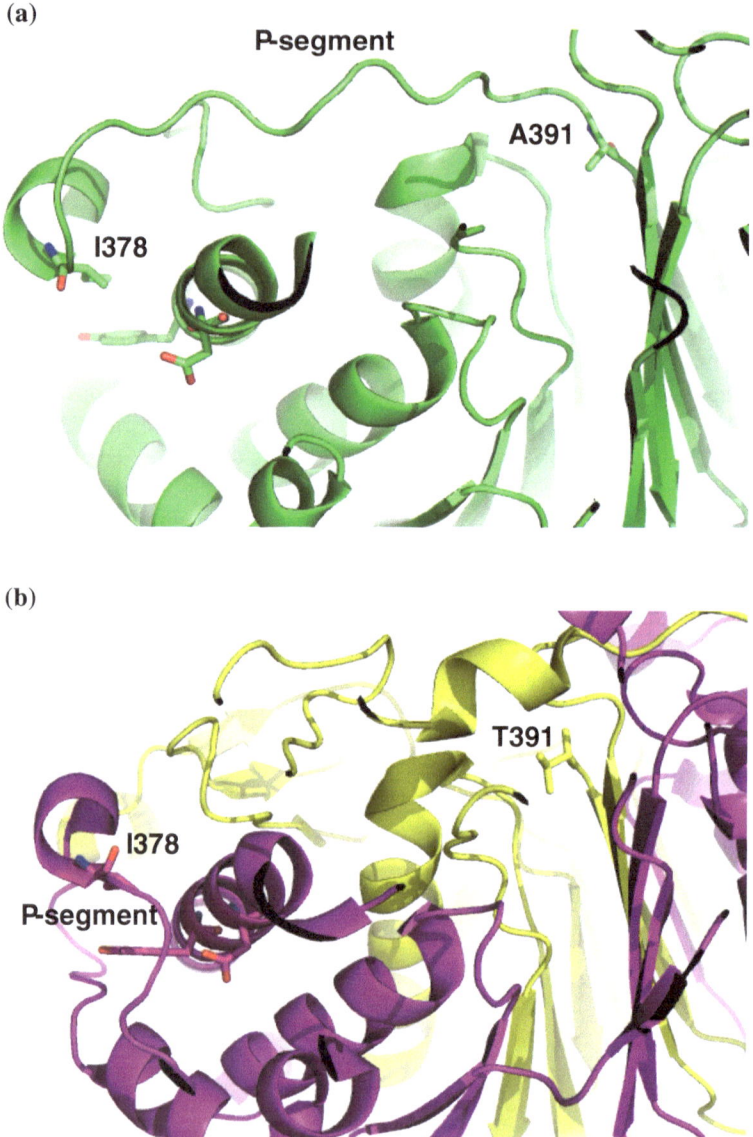

Fig. 9 Structures of inactive precursor (**a**) and mature form of EcGT and (**b**). In the mature enzyme, the large subunit is shown in *violet* and the small subunit in *yellow*. The position of catalytic important Thr (Thr391), Ile378, P-segment, and lid loop have been evidenced. In the precursor, the Thr is replaced by an Ala. To emphasize the similar orientation of the structures in the two panels, the side chains of Tyr332 and Asp336 are also shown

this segment in the large subunit of the protein from *Bacillus subtilis* is not significantly affected by the autocatalytic processing, being located close to the N-terminal region of the small subunit [40].

4.2 Reaction Mechanism

γ-GTs catalyze the cleavage of the γ-glutamyl linkage of γ-glutamyl compounds, such as GSH and its S-conjugated, including leukotriene C4, S-nitroso-glutathione (GSNO), and GSH adducts of xenobiotics formed by the action of glutathione-S-transferases [8]. In a second step, most γ-GTs catalyze the transfer of the γ-glutamyl moiety to other aminoacids or short peptides [54].

In the most widely accepted reaction mechanism, the highly conserved Thr, which is also responsible for autoprocessing, attacks the C=O group of the γ-glutamyl-compound to form a γ-glutamyl-enzyme intermediate. The intermediate then reacts with water, to release glutamate in a hydrolysis reaction, or with an acceptor, to give a transpeptidation reaction forming new γ-glutamyl compounds (Fig. 10). The simplest acceptor substrates include single amino acids and glycine dipeptides (Gly-Ala or Z-Gly, where Z is W, F, Y, H, A, Q, C, M, S, L, K, A, or G).

The attack is likely promoted by one or more residues adjacent to the catalytic Thr, which can form hydrogen bonds with the catalytic Thr, increasing the reactivity of its hydroxyl group that then attacks the γ-glutamyl peptide [23]. Barycki and coworkers have proposed that the hydroxyl group of a second Thr, which follows the catalytic one in the sequence of the small subunit and is well conserved in the γ-GT family, forms two hydrogen bonds with the catalytic Thr [47]. This Thr–Thr dyad is critical for efficient cleavage of the γ-glutamyl peptide bond of GSH. Recombinant DNA techniques have been used to produce BlGT and HpGT variants in which either the catalytic Thr or the second Thr of the dyad are replaced with an alternative residue. These studies have shown that the former Thr is essential for the catalytic activity, since most of unprocessed proteins, in which this residue was mutated in Ala, were found almost completely inactive [39, 47], while the T → S substitution results in a protein with catalytic activity significantly altered when compared to the native enzymes. Notably, the replacement of the latter Thr of the dyad with lysine can impair the autoprocessing [55]. To establish whether the lack of catalytic activity by the T → A mutants is due to the unprocessing of the enzyme or to the absence of the hydroxyl group of the Thr, a co-expression system in which the processing of the enzyme was uncoupled from the catalytic activity was studied. Co-expression of the small and large subunits results in a catalytically competent enzyme [47].

4.3 γ-Glutamyl Binding Site

As described above, the reaction catalyzed by γ-GTs proceeds via the formation of a γ-glutamyl-enzyme intermediate followed by nucleophilic substitution by water, amino acids, or peptides. Formation of the intermediate was demonstrated by a number of experimental evidences acquired by chemical, kinetic, and crystallographic studies. A clear picture of the γ-glutamyl-enzyme intermediate was

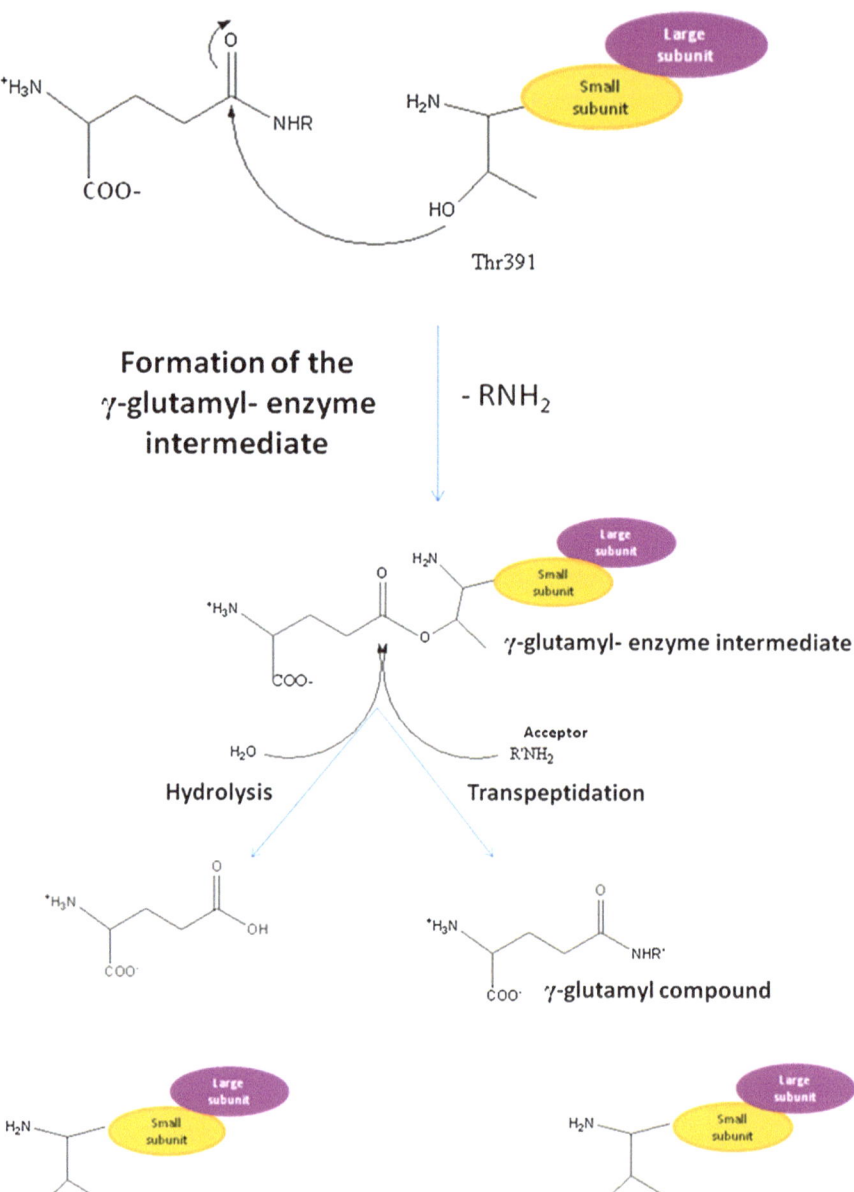

Fig. 10 Proposed mechanism for catalysis of γ-glutamyl peptide cleavage by γ-GTs. In mature enzyme, the hydroxyl group of the N-terminal Thr residue of the small subunit, conserved in all homologous γ-GTs, attacks the γ-glutamyl peptide bond of GSH and leads to the formation of a tetrahedral transition state, whose collapse leads to the formation of a tetrahedral intermediate (γ-glutamyl enzyme complex). The intermediate is stabilized through interactions with two conserved glycines, and concomitant expulsion of the leaving group. A second Thr in the sequence of the small subunit can increase the reactivity of catalytic Thr. Adapted from Castellano and Merlino [102]

obtained solving the structures of EcGT in complex with GSH (γ-glutamyl-enzyme intermediate, PDB code 2DG5) and L-glutamate (PDB code 2DBX) (Fig. 11). The substrate-binding pocket is located at the bottom of a deep groove, close to the catalytic Thr (γ-glutamyl binding site) (Fig. 11). This site exhibits a scarce specificity, since it is able to accept both L- and D isomers of glutamic acid, and a number of different substrates.

When bound within γ-glutamyl binding site, the carbonyl group of the γ-glutamyl moiety is covalently linked to the OG atom of the catalytic Thr. The γ-glutamyl moiety is held in position by an extensive network of hydrogen bonds and salt bridges with several residues (Fig. 11). In particular, in the complex of EcGT with L-glutamate, the carboxyl group of the ligand is bonded with R114, S462, and S463, whereas the amino group interacts with N411, Q430, and D433 (Fig. 11b). The γ-glutamyl carbonyl oxygen is hydrogen bonded with the main-chain atoms of G483 and G484. Other known active site residues are L461, P482, and I487. Except for R114, all residues involved in γ-glutamyl group recognition belong to the small subunit. Mutational studies of hGT have supported the involvement of conserved Asp and Ser residues (D423, S451, and S452 in hGT) in the γ-glutamyl binding site formation [56, 57].

It has been proposed that when the γ-glutamyl binding site is occupied by a substrate (or inhibitor) the lid loop has a well-defined position and shields the catalytic pocket from the solvent, otherwise, when the pocket is empty, the lid loop is disordered. Thus, the lid loop, which is conserved in all eukaryotic γ-GTs, should regulate the access of the substrate to the active site or the binding of the substrate to the active site cleft. Multiple sequence alignments showed that several microbial enzymes including GthGT and those from *Bacillus subtilis, Bacillus halodurans, Thermoplasma acidophilum, Deinococcus radiodurans, Picrophilus torridus*, and *Thermus thermophilus* lack the sequences corresponding to this loop [46, 58]. In EcGT, Y444, located at the middle of this loop, is hydrogen-bonded with N411, thus forming the wall of the substrate binding pocket. Y444 acts as a gating residue. Molecular dynamics studies have revealed that in hGT the lid loop (residues 423–438) exhibits large fluctuations in both the apo and the holo state, hence also in the presence of the γ-glutamyl moiety [59]. These results undermine previous hypothesis on the reduction of mobility of the loop in EcGT and HpGT upon the formation of the covalent adduct. In hGT, the gating residue is replaced by a Phe, which lacking the side chain OH group, cannot form the above described stabilizing interaction. As a result, the lid loop of hGT in the substrate-bound complex should exhibit larger fluctuations than those of the corresponding region of EcGT and HpGT. The higher conformational variability of hGT lid loop, that is, the ability to change its structure from an open to a closed conformation, could explain higher transpeptidase activity (faster product release) of hGT when compared to the other members of the family.

Interestingly, the structure of the L-glutamate-bound BsGT revealed that neither the lid loop nor alternative ordered segments cover the active site, which remains exposed to the solvent [40]. γ-GTs with solvent exposed active sites could accept other potential substrates.

(a)

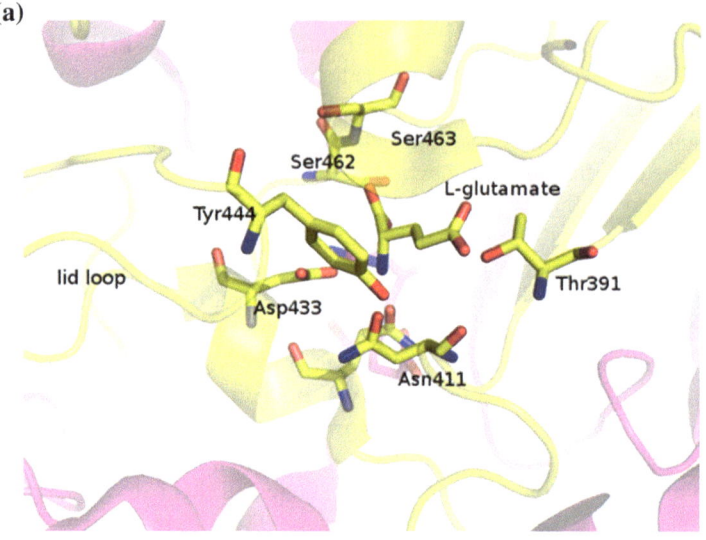

(b)

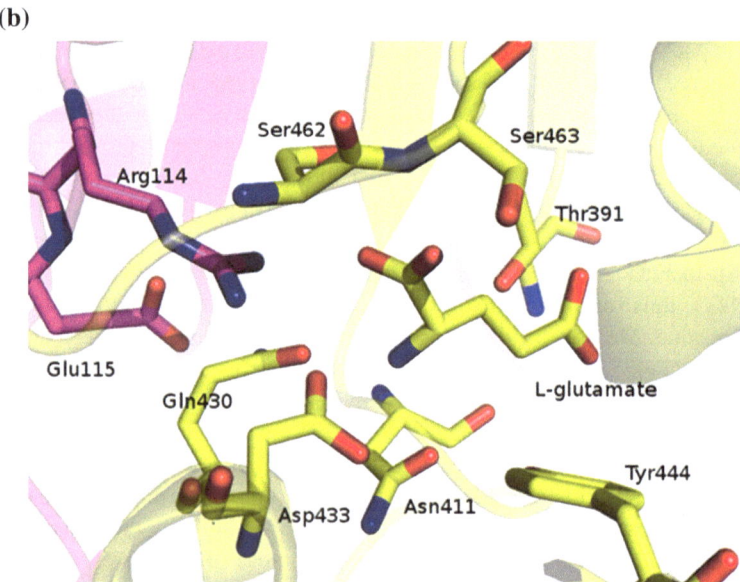

Fig. 11 a γ-glutamyl binding site of EcGT. The large subunit is shown in *violet* and the small subunit in *yellow*. **b** Details of the γ-glutamyl binding site. The position of the L-glutamate, of the catalytic Thr and of Tyr444 is evidenced

4.4 Acceptor Binding Site

Structural data aimed at identifying residues involved in the recognition of the acceptor and forming the acceptor binding site are still missing. Some results indicate that the acceptor binds to the site occupied by the leaving group [60, 61]. This hypothesis would suggest that the catalytic mechanism proceeds in a sequential manner, that is, the enzyme binds the γ-glutamyl compound and the acceptor in the same site in a sequential way. However, at the moment, this hypothesis is not conclusive. Further attempts to unveil the molecular bases of acceptor recognition are based on docking studies performed using structural models of glycyl-glycine and *Picrophilus torridus* γ-GT (PtGT) or its Y327N mutant [62] and on site directed mutagenesis and computational studies performed on hGT and on CapD from *B. anthracis,* a protein related to γ-GTs but missing the ability to hydrolyze GSH [59]. A computational study performed on the PtGT Y327N mutant has suggested that the glycyl-glycine docks perfectly in a binding pocket where five residues (L87, E90, Y305, N327, and S348) can form hydrogen bonds with the ligand [62]. In the same binding pocket, in the case of wild-type PtGT, only four residues (S348, F350, T351, and Y425) interact with the acceptor. This result is strengthened by comparative intrinsic fluorescence emission spectra. In fact, a shift of maximum emission wavelength from 330 to 345 nm is observed when the acceptor-free mutant Y327N is compared to the same protein in the presence of glycyl-glycine. In the case of the wild type, addition of the acceptor causes only a fluorescence quenching.

The computational studies performed on hGT have suggested that the putative acceptor binding site is located in the region along the groove of the donor binding site. The binding pocket in human enzyme should be formed by a number of polar residues such as D46, N79, H81, S82, Y403, Q476, and K562. Site-directed mutagenesis studies of hGT and CapD identify residue K562 in hGT (R520 in CapD), N373, and R432 in CapD (R in EcGT) as involved in the formation of the acceptor site. It is worth noting that two small loops of hGT at the acceptor site, formed by residues 459–464 and residues 549–553 were found to exhibit large motions in molecular dynamics studies [59].

Since the mechanism of transpeptidation has not been clarified yet, further studies are needed to verify the speculative aspects of these models on the basis of new structural insights. These studies should also explain why the absence of any single glycosylation in mammalian enzymes affects both the transpeptidation and hydrolysis reactions of hGT [34].

4.5 Transpeptidation Versus Hydrolysis

γ-GTs catalyze both hydrolysis and transpeptidation reactions. In vivo, the primary reaction is the hydrolysis [34]. The two reactions have different pH-rate profiles.

Their behavior can also be different when homologous enzymes are compared. For example, transpeptidation shows a bell-shaped pH-rate profile for hGT, but only an ascending limb in EcGT [25]. The hydrolysis reaction yields a relatively flat pH-rate profile. These differences make a γ-GT able to catalyze one of the two reactions selectively, by adjusting the pH of the reaction mixture. Using rat γ-GT it has been shown that the formation of acyl-enzyme intermediate is the rate-limiting step [54, 63, 64].

Interestingly, some γ-GTs from extremophilic organisms are specialized in hydrolase activity [46, 58]. Comparative sequence and structural analysis between extremophilic γ-GTs and mesophilic counterparts with high sequence identity suggested that the transpeptidation is mediated by residues S462 and N411 (numbering scheme of *E. coli*), which are located in γ-glutamyl binding site (Fig. 11). In site-directed mutagenesis studies, in which these residues were introduced in the sequence of the hydrolase specialized γ-GT from *Picrophilus torridus*, the enzyme acquired significant transpeptidase activity [62].

4.6 Enzymatic Assays

γ-GT activity has been assessed by using a number of different substrates. Determination of the hydrolytic activity has been generally performed using γ-glutamyl-*p*-nitroanalide (GpNA) [31, 65]. The release of *p*-nitroaniline (pNA) was monitored by spectrophotometer at 410–412 nm ($\varepsilon = 8800$ M^{-1}). When the activity of the recombinant proteins is too low to allow continuous monitoring, the release of pNA is assessed by end point assay after 5–10 min of incubation with the substrate. When high concentrations are required, the limited GpNA solubility is overcome by using γ-glutamyl-(m-carboxyl)-*p*-nitroanilide. The transfer of the γ-glutamyl group to a dipeptide acceptor, which occurs by a modified Ping-Pong mechanism, has been generally followed using GpNA and high concentrations of the acceptor glycyl-glycine (GlyGly), at pH 8.0 or higher (Fig. 12).

Recently, Hanigan and coworkers have developed a new quantitative assay to measure hydrolytic activity of γ-GTs [66]. According to this new method, the

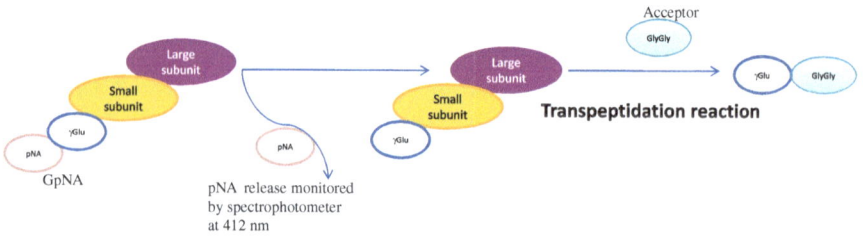

Fig. 12 Standard γ-GT enzymatic assay

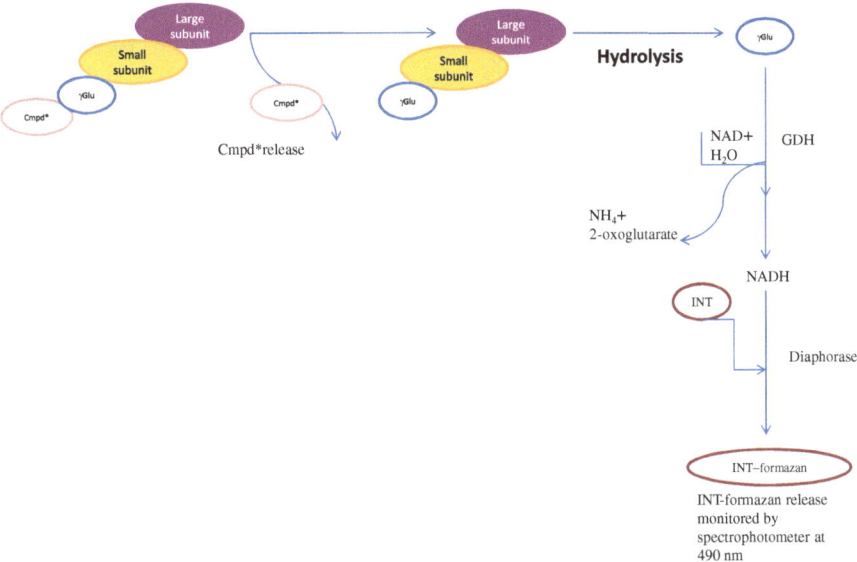

Fig. 13 Schematic representation of the new quantitative γ-GT enzymatic assay developed by Hanigan and coworkers [66]

hydrolysis of γ-glutamyl bonds is detected by a coupled assay in which the glutamate released by the hydrolysis of the γ-glutamyl bond is quantified by a modified version of an assay developed by Beutler [67]. In particular, the glutamate is oxidized by glutamate dehydrogenase (GDH), reducing NAD^+ to NADH. NADH in the presence of diaphorase reduces iodonitrotetrazolium (INT) to INT-formazan, a colored product, which can be detected at 490 nm. The production of glutamate in the second reaction can be monitored continuously. In the third reactions, INT-formazan is produced in stoichiometric ratio with the amount of L-glutamate (Fig. 13).

4.7 Biochemical Properties of γ-GTs

Despite the considerable sequence identity among members of the γ-GT family, a direct comparison of the catalytic activity of enzymes from different source is difficult, because of the different experimental conditions and substrate molecules used to study each enzyme. However, a rough comparison indicates that significant catalytic differences exist among γ-GTs. In general, plant γ-GTs are similar to the mammalian enzymes in their biochemical characteristics [68], whereas bacterial γ-GTs are generally catalytically less powerful than their eukaryotic homologs.

For example, the activity of rat kidney γ-GT is about 100-fold higher than that of EcGT [56]. hGT catalyzes transpeptidation 180-fold faster than hydrolysis, while HpGT shows only a modest 2-fold increase [47]. Large differences in catalytic activities also exist when bacterial homologs are compared. BsGT is 33-fold more active than EcGT [55]; furthermore, as already mentioned, extremophilic bacterial γ-GTs display reduced hydrolase activities, when compared to the other bacterial and eukaryal counterparts, and do not display any ability to transfer the γ-glutamyl group to acceptors [46, 58]. Further differences occur for the catalytic rate. Eukaryotic γ-GTs can be stimulated up to 100-fold by acceptors, whereas pro-karyotic enzymes only exhibit marginal changes: about 2-fold in EcGT and about 10-fold in BsGT [25, 26, 56].

For the reactions catalyzed by γ-GTs, the optimal pH is generally between 8 and 9, with the exceptions of the enzymes from *Bacillus subtilis* SK 11.004 and *Pseudomonas nitroreducens* which have an optimum pH of the transfer reaction of 10 and 10.5, respectively [69, 70]. The optimal temperatures for γ-GT activities range from 37 to 60 °C. Some enzymes are highly stable below 50 °C [70], whereas others are highly sensitive to thermal inactivation [26, 71]. For example, the enzyme from *Bacillus pumilus* KS is highly thermostable, retaining 50 % of the original activity at 70 °C [72], whereas GthGT and BlGT show approximately 80 and 40 % of the original catalytic activity at 45 °C [44]. Altogether, these features could reflect different mechanisms of adaptation of the enzyme to colonize different niches. Otherwise, they could remark different evolutionary relationships in γ-GT family, with extremophilic γ-GTs as the ancient progenitors [58].

Since γ-GT has been considered a suitable enzyme in industrial processes that require high-salt conditions, such as the manufacture of fermented foods where >18 % NaCl is generally used, it is important to find halotolerant γ-GTs that retain complete transpeptidase activity in the presence of such high concentration of NaCl. At the moment, very few halotolerant γ-GTs are known; the majority of the members of the family are inactivated in the presence of salt. Examples of halo-tolerant γ-GTs include BsGT, BlGT, GthGT, and a monomeric 30 kDa γ-GT purified from *B. licheniformis*, probably generated by proteolytic digestion of the mature BlGT by subtilisin [33, 73–76]. Hypotheses on the structural basis for the salt tolerance of BsGT and GthGT have been reported [76]. The peculiar feature of halotolerant γ-GTs to be stable in the presence of large amount of salts is related to the existence of a higher proportion of acid residues on their surface with respect to their non-halotolerant counterparts. These acid patches on the surface may allow proteins to remain in the hydrated state and avoid self-aggregation even under high-salt conditions [40]. In this respect, GthGT can be a prototype. In view of its thermohilic origin and halophilic behavior, GthGT possesses a larger ratio of negatively charged (D + E) residues to positively charged (R + K) ones and a larger percentage of exposed negatively charged side-chains with respect to mesophilic counterparts. These side chains are globally distributed on GthGT molecular surface. This finding is in line with the results of several independent

Table 3 Effects of mutations on enzymatic activity of hGT

Human GT	Mutated residue	Effect on catalytic activity
	K100 → N:	No effect on activity
	E102 → Q	No effect on activity
	R107 → K	Reduces enzyme activity by 99 %
	R107 → Q	Abolishes enzyme activity
	R107 → H	Abolishes enzyme activity
	E108 → Q	Reduces enzyme activity by 98 %
	R112 → Q	No effect on activity
	R139 → Q	No effect on activity
	R147 → Q	No effect on activity
	R150 → Q:	No effect on activity
	H383 → A	Reduces enzyme activity by 66 %
	S385 → A	No effect on activity
	S413 → A	No effect on activity
	D422 → A	Reduces enzyme activity by 90 %
	D423 → A	Abolishes enzyme activity. Increases KM by over 1000-fold
	S425 → A	No effect on activity
	S451 → A	Reduces enzyme activity by 99 %. Abolishes activity when associated with S452A
	S452 → A:	Reduces enzyme activity by 99 %. Abolishes activity when associated with S451A
	C454 → A	No effect on activity
	H505 → A	Reduces enzyme activity by 90 %

studies suggesting that halophilic enzymes present higher proportion of acid residues on the surface than their non-halophilic homologs [77, 78] and with a recent comparative structural analysis of residue distribution in a database comprising 15 pairs of halophilic/non-halophilic homologs [76]. On the bases of these results, γ-GT mutants with increased resistance versus high salt environment could be designed.

Several molecules are able to uncouple hydrolysis and transpeptidase activity of γ-GTs. Activators of the hydrolase activity include maleate, free bile acids cholate, chenodeoxycholate, and deoxycholate, and some metal ions. Maleate stimulates the hydrolase activity of rat kidney γ-GT [79], but inhibits its transpeptidase activity. Hippurate and its derivatives are also capable of uncoupling hydrolysis and transpeptidase activity [80], whereas free bile acids affect the kinetics of both hydrolysis and transpeptidation [81]. Regarding metal ions as activators, Li^+, Rb^+, K^+, Na^+, Cs^+, Mg^{2+}, Ca^{2+}, Co^{2+}, and Mn^{2+} activate EcGT [82], whereas the enzymatic activity of BsGT is enhanced in the presence of Al^{3+}, Mg^{2+}, K^+, Na^+ [70]. Mg^{2+}, K^+, and Na^+ also enhance catalytic activity of BlGT [44].

Site directed mutagenesis has been used to reveal which residues are important for the catalytic activity of γ-GTs. Mutations investigated in the human enzyme are reported in Table 3.

Fig. 14 Structures of
glutamine analogs that inhibit
γ-GTs: Azaserine and
Acivicin

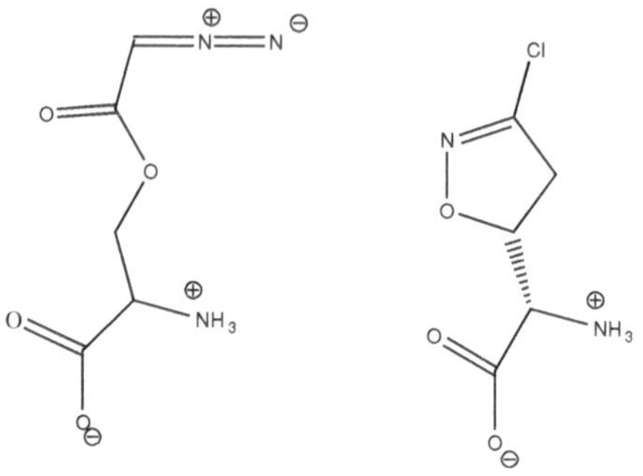

Azaserine   Acivicin

4.8 Quest for Inhibitors

Rational design of γ-GT inhibitors has encountered a lot of difficulties, due to uncertainty about γ-GT reaction mechanisms. Compounds which are known to inhibit γ-GT include the glutamine analogs Acivicin (L-(αS,5S)-α-amino-3-chloro-4,5-dihydro-5-isoxazoleacetic acid), 6-diazo-5-oxo-L-norleucine (DON), and Azaserine (O-diazoacetyl-L-serine) [83] (Fig. 14).

These inhibitors have been widely used for the in vitro and in vivo experiments on γ-GTs. Acivicin, the most potent inhibitor that has been tested clinically, has a protective effect on cisplatin-induced nephrotoxicity [84] and suppresses γ-GT-dependent oxidative damage in ischemic rat kidney [85]. However, this molecule also inhibits various glutamine amidotransferases, including imidazole glycerol phosphate synthase and guanine monophosphate synthetase, and inactivates a number of biosynthetic enzymes for purine and pyrimidine, amino acids, and amino sugars, resulting in a potent cytotoxicity [86–88]. Both Acivicin and Azaserine cause bone marrow suppression.

Design of γ-GT inhibitors based on studies of active site has led to the identification of additional γ-glutamyl analogs. Lherbet and Keillor have designed sulfur derivatives of L-glutamic acid that inhibit γ-GTs [89] and are less toxic than the glutamine analogs. Han and co-workers have synthesized and tested a series of γ-(monophenyl)phosphono glutamate analogs which also function as inhibitors of γ-GTs [90]. Among these molecules, 2-amino-4-(3-(carboxymethyl)phenyl(methyl) phosphono)butanoic acid (GGsTOP, Fig. 15) seems one of the most promising.

GGsTop is an electrophilic phosphonate phenyl ester, it is non-toxic but not stable enough to be used for in vivo studies [91]. It does not inhibit the asparagine synthetase and shows 100-fold higher inhibitory activity toward hGT than that of Acivicin [90, 91].

Another potent inhibitor is the γ-boronic acid analog of L-glutamic acid, 3-amino-3-carboxypropaneboronic acid (γ-boroGlu, Fig. 16) [92]. Weak inhibitors are alkylboronic acids [93], m-formylphenylboronate [93], and the serineboric acid complex [94].

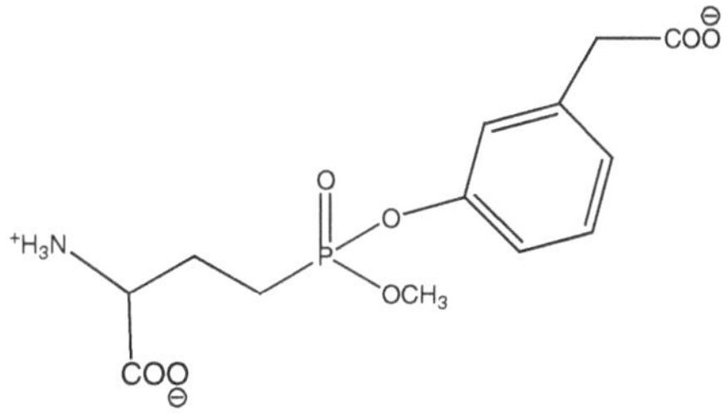

Fig. 15 Structure of
GGsTOP

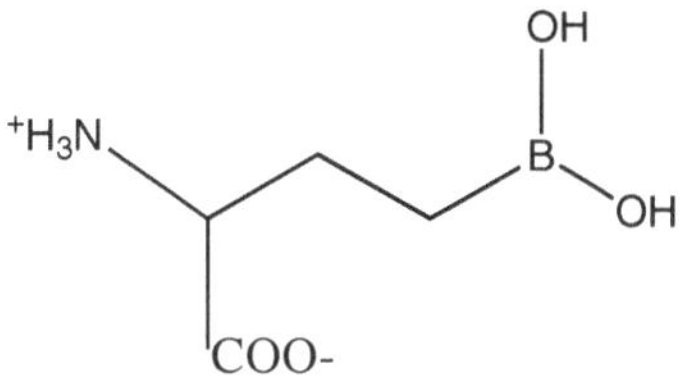

Fig. 16 Structure of
γ-boroGlu

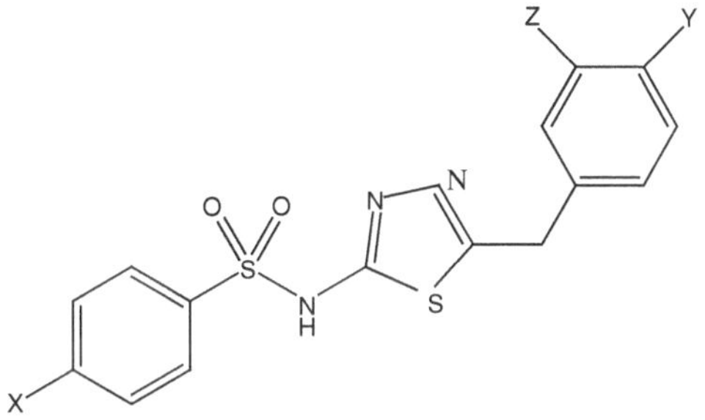

Fig. 17 Structure of analogs
of OU749. X, Y, Z may be,
for example, H, Cl, F, Br, I,
OH, an alkoxy, or NO$_2$ [95]

Recently, a novel class of γ-GTs uncompetitive inhibitors that are structurally distinct from and less toxic than the glutamine analogs has been also developed. These compounds, which are derived from N-[5-(4-methoxybenzyl)-1,3,4-thiadiazol-2-yl] benzenesulfonamide, called as analogs of OU749 (Fig. 17) [95], should not occupy the γ-glutamyl binding site, but the acceptor binding one (see below). These inhibitors are species-specific since they inhibit hGT, but not mouse, rat, and pig enzymes [96].

4.9 Inhibitor Mechanisms

Inhibition mechanisms of Acivicin and Azaserine have been recently elucidated by X-ray crystallographic studies [97]. CA atom superposition of the Acivicin and Azaserine complexes (PDB accession codes: 2Z8I and 2Z8K) with the ligand-free enzyme structure (PDB accession code: 2DBU) yields a RMSD <0.2 Å, indicating that binding of these inhibitors does not significantly affect the overall structure of

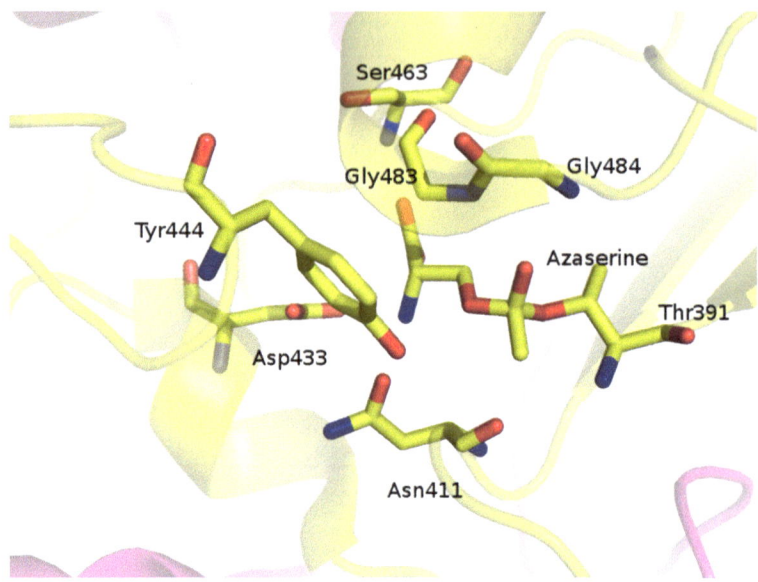

Fig. 18 Binding mode of Azaserine. Location and orientation of the molecules are very similar to those of the γ-glutamyl enzyme intermediate

the (mature) protein. X-ray structures of the complexes EcGT-Acivicin and EcGT-Azaserine have disclosed that the tight protein-inhibitor association is largely due to the formation of an adduct with γ-GTs, which is structurally analogous to transition state that forms during the reaction pathway for catalysis (Fig. 18). In particular, these inhibitors form a covalent bond with the OG atom of the catalytic threonine. However, structure and hybridization of these inhibitors are unusual. In EcGT-Azaserine complex, the CH atom of Azaserine (Fig. 18) involved in the formation of the covalent bond with the Thr is hybridized sp^3 and adopts a tetrahedral geometry (Figs. 18, 19). This is not the final inactivation product: it rearranges to form a more stable adduct that irreversibly inhibits the protein. A proposed inhibition mechanism of Azaserine is reported in Fig. 19. In the case of EcGT-Acivicin complex, the imino carbon of the dihydroiaoxazole ring is linked to the catalytic Thr. Interestingly, the inspection of the electron density maps of this adduct reveals that Cl atoms are removed from the inhibitor upon the binding (Fig. 20). The proposed inhibition mechanism of Acivicin is reported in Fig. 21.

The inhibition mechanism of the other known γ-GT inhibitors has not been confirmed by experimentally determined structures of adducts, but have been supposed on the bases of biochemical data. The inhibition mechanism supposed for GGsTOP is reported in Fig. 22. Also in this case, the inhibitor should link to the catalytic Thr in the enzyme active site.

γ-boroGlu (Fig. 23) [94] and DON [98] should form a covalent, tetrahedral adduct with the protein as well.

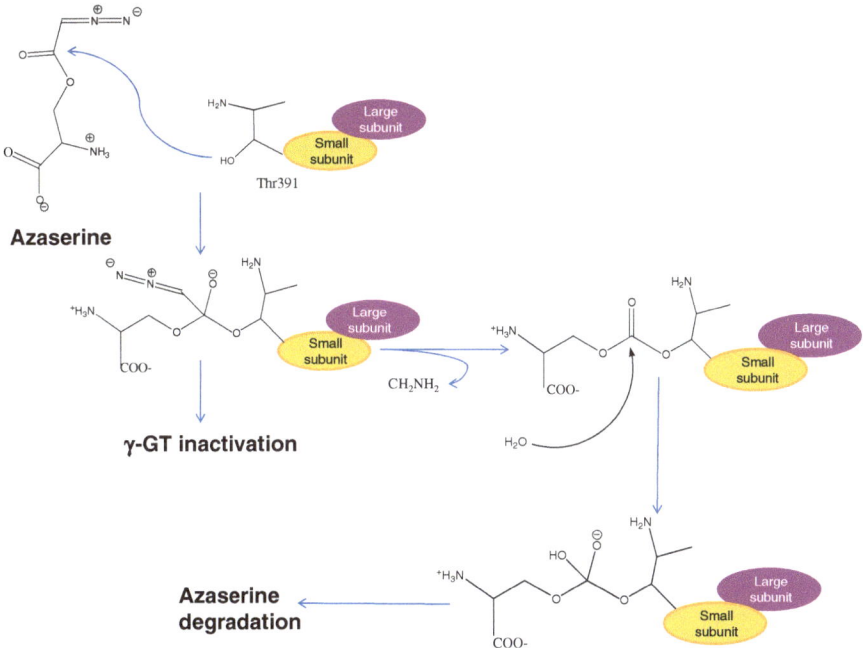

Fig. 19 Proposed inhibition mechanism for Azaserine. The OG atom of catalytic Thr attacks the carbonyl carbon of Azaserine leading to the formation of a tetrahedral adduct. The activated diazoketone may inactivate the enzyme by alkylation or eliminate diazomethane to form an *O*-carboxyseryl enzyme intermediate. The addition of a water molecule to the carbonyl forms a tetrahedral adduct that may lead to the overall degradation of Azaserine by the enzyme

On the contrary, OU749 mechanism of action seems different. The predicted binding model of OU749 with hGT showed that this molecule is located in the acceptor binding site with its phenylcarboxylate group that can form a hydrogen bond with Lys562 [59]. The discovery of OU749 is promising for the design of non-toxic, species-selective inhibitors of γ-GTs directed to the acceptor binding site.

4.10 Conformational Stability of γ-GTs

Understanding the structure-stability-function relationship of an enzyme is crucial for its applicative aspects. Conformational studies may provide insight into the molecular basis of the enzyme stability, which in turn can be used to design protein variants with special properties for biotechnological applications. Few reports deal with biophysical characterization of precursors and mature forms of γ-GTs. This is in part due to the difficulty to purify γ-GTs at homogeneity, since a mixture of the

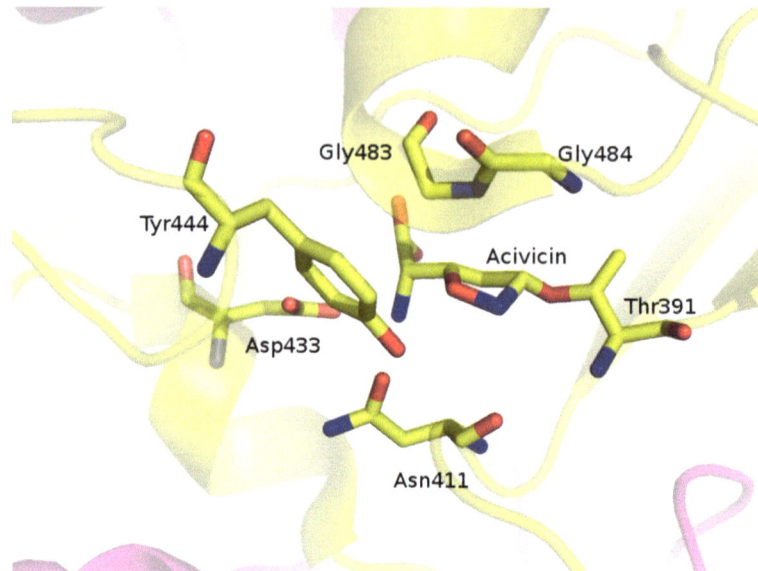

Fig. 20 Binding mode of Acivicin. Location and orientation of the molecule is very similar to that of the γ-glutamyl enzyme intermediate

unprocessed and mature protein is often obtained [46, 58, 72]. Unfolding analyses using circular dichroism and tryptophan emission fluorescence have revealed that members of the γ-GT family display different sensitivity toward temperature and guanidinium hydrochloride induced-denaturation [75, 99, 100]. Thermal denaturation of EcGT and BlGT follows a simple irreversible two-state mechanism (N → U) [75], whereas that of GthGT was described using a three-state model involving the formation of a stable intermediate [100]. Among the characterized enzymes, GthGT is the most thermostable, in line with its thermophilic origin, whereas BlGT is the most resistant to chemical denaturation with guanidinium hydrochloride. Interestingly, it has been recently shown that heat-inactivated EcGT left at 4 °C temperature recovers its activity, whereas heat-treated BsGT does not. These results have been explained with the hypothesis that heat-treated EcGT dissociates into large and small subunits, which reassemble to form a catalytically competent enzyme at low temperature [101].

4.11 Search for Thermostable γ-GTs

In recent years, demands for thermostable enzymes have increased in industrial processes. Biocatalyst thermostability allows a higher operation temperature, which

Fig. 21 Proposed inhibition mechanism for Acivicin. Catalytic Thr attacks the C3 imino carbon in the dihydroisoxazole ring of Acivicin. This results in the formation of anaza-anion. The O–N bond is then cleaved with concomitant loss of the hydrogen in C5 and of the Cl atom to give a β-oxoimidoyl group. Finally, the imidoyl carbon attached to the OG atom of Thr becomes sp^3 hybridized by elimination of the H in C4 with concomitant ring closure and formation of a double bond between C4 and C5

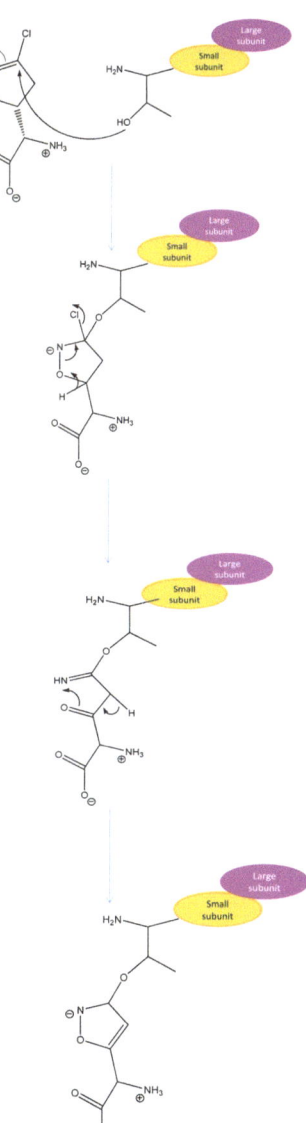

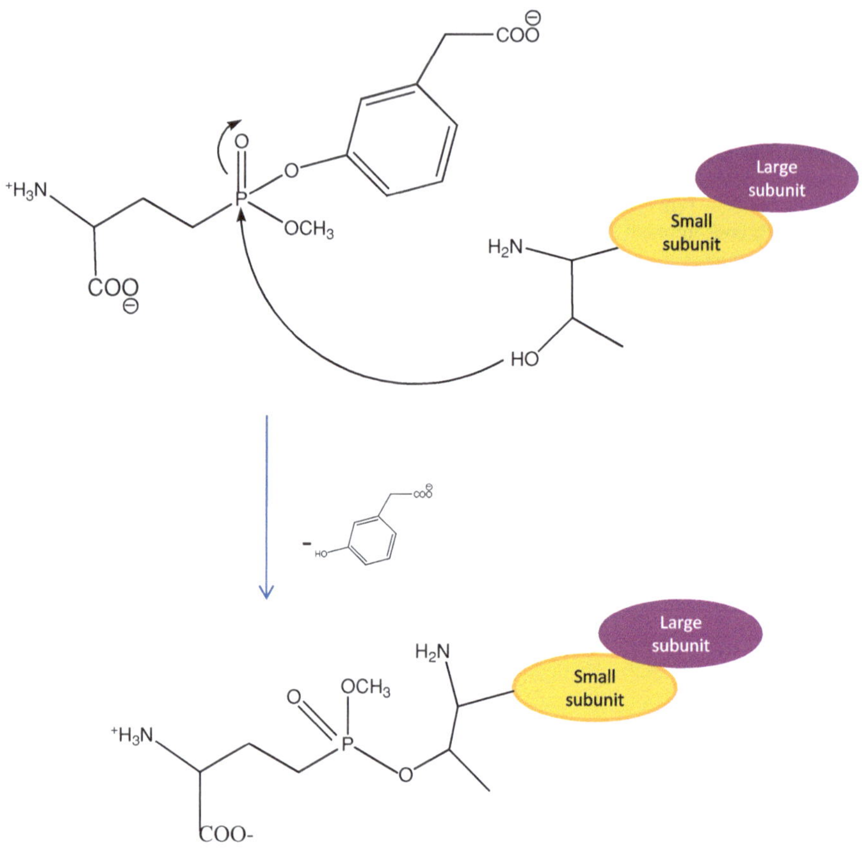

Fig. 22 Proposed inhibition mechanism for GGsTOP

is clearly advantageous because of a higher reactivity (higher reaction rate, lower diffusion), higher process yield (increased solubility of substrates and products and favorable equilibrium displacement in endothermic reactions), lower viscosity and fewer contamination problems. For this reason, the purification and characterization of thermostable γ-GTs is required. Thermophiles represent an obvious source of thermostable enzymes. Recent progress in genomics has provided a wealth of genome sequences, and several sensitive homology search programs have discovered γ-GT homologs in many thermophilic microorganisms [62].

Cloning, expression, and purification of thermostable γ-GT from *G. thermodenitrificans* was reported in 2010 [46]. GthGT was the first thermostable γ-GT to be characterized. One year later, overexpression of the gene encoding the γ-GT from *T. thermophilus and D. radiodurans* was reported [102]. More recently, the protein from *P. torridus*, an acido-thermophilicarcheon, has been characterized [62].

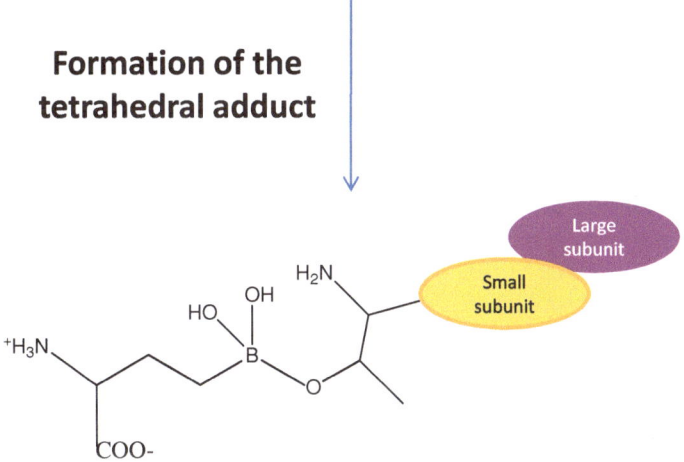

γ-boroGlu

**Formation of the
tetrahedral adduct**

Fig. 23 Proposed inhibition mechanism for γ-boroGlu [92]

5 γ-GT in Bioclinical

hGT levels in blood are routinely measured in clinical laboratories. Indeed, γ-GT
serum is mainly derived from liver, thus the enzyme is used as marker of liver or
biliary tract-associated diseases [103, 104]. The higher the γ-GT level the greater
the "insult" to the liver. Serum levels of γ-GT are affected by many factors:
alcohol intake [13], body fat content, plasma lipid/lipoproteins, and glucose levels.

Increased serum γ-GT levels have also been associated with pancreatitis, type II
diabetes [105], cardiovascular disease [106, 107], stroke, and hypertension [108].
However, the reasons of these correlations are not always known. In patients with
hypertension, γ-GT is positively associated with inflammation markers such as
fibrinogen, C-reactive protein, and F2-isoprostanes[109] and is expressed in human
atherosclerotic lesions, thus contributing to the progression of atherosclerosis
[109]. Finally, γ-GT plays a key role in the production of free radical species
through its interaction with iron [110]. Therefore, elevated γ-GT levels could be a

marker of inflammation condition and oxidative stress, typical features of hypertension. An increase in serum levels of γ-GT has been also suggested to be predictive of atherosclerosis [111], but the mechanism of this association has not been elucidated. One hypothesis is that γ-GT acts as proatherogenic factor [112]. Otherwise, the increase of γ-GT could be a marker of the presence of the metabolic syndrome [113]. Indeed, evaluation of γ-GT activity in obesity is associated with insulin resistance and consequent metabolic syndrome [114].

The reference range for γ-GT is 0–30 IU/L. Levels of γ-GT are higher in men than in women (ca 25 %) and, in women, increase with age [115]. Infants levels overcome 6–7 times the upper limit of the adult reference range, which declines to adult levels around age 7 months [116]. Elevated γ-GT levels have also been associated with use of many drugs, including barbiturates, antibiotics, histamine receptor blockers, antifungal agents, antidepressants, and testosterone. Smoking can also increase γ-GT levels. Clofibrate and oral contraceptives can decrease γ-GT levels.

γ-GT has been implicated in many physiological disorders, such as neurodegenerative diseases [117, 118]. Moreover, γ-GT has also been reported as a virulence factor associated with the colonization of the gastric mucosa by *H. pylori*, the pathogen responsible for gastritis, ulcer and gastric cancer [27].

γ-GT deficiency, a rare autosomal recessive disease documented in five patients [114] has been linked to disruptive glutathione homeostasis, DNA damage, reproductive defects, mental retardation, and cataract [119]. Genetic γ-GT deficiency in mouse impairs glutathione metabolism and transport and results in systemic glutathionemia, glutathionuria, and oxidant stress in the kidney [120].

5.1 Pathways of γ-GT Induction

The appearance of γ-GT-positive foci in animals exposed to chemical carcinogens and the increased expression of γ-GT in proliferating pre-neoplastic foci in the liver have suggested that γ-GT can be considered as an early marker of neoplastic transformation [5, 8, 13, 121, 122]. It has been shown that γ-GT is up-regulated in different cell types after acute exposure to oxidative stress [123] also involving the extracellular signal-regulated kinase (ERK) and p38 mitogen-activated protein kinase (MAPK) pathways. Indeed, γ-GT expression and Ras-MAPK pathways activation have been correlated in colon cancer cells following γ-irradiation [124], as well as exposure to oxidative stress [125]. Reactive oxygen species (ROS) have been implicated in the process of carcinogenesis, and, the redox regulation of many genes in response to ROS/electrophiles seems to modulate γ-GT expression.

Interestingly, γ-GT mRNA was shown to be induced also by cytokines, including tumor necrosis factor alpha (TNF-alpha) [126], and interferon (IFN)-alpha and -beta [127, 128]. These results suggest that the increased γ-GT expression could reflect in some conditions a defensive mechanism against oxidative stress or in other conditions a regulatory mechanism, involving metabolism of leukotrienes and GSNO.

5.2 γ-GT and Cancer

Increased levels of γ-GT have been observed in cancer of ovary, colon, liver, astrocyticglioma, soft tissue sarcoma, melanoma, and leukemias [129]. Elevated γ-GT activity has also been found in melanoma cells with an increased invasive growth, and γ-GT expression with unfavorable prognostic signs has been found in human breast cancer. In contrast, in prostatic, colorectal and breast cancer any correlation between expression and standard clinical pathological parameters have been found. These differences can be explained by considering the high variability of cancer cells and the effect of other factors, such as the environment, drugs and diet, which may alter the phenotype of neoplastic lesions [129].

Several studies have addressed the relationships of γ-GT activity with the malignant phenotype [13]. The involvement of γ-GT in cellular supply of extracellular GSH and the increased resistance to pro-oxidant drugs observed in several γ-GT-expressing cell lines have suggested that γ-GT plays a role in the cellular defense mechanisms. On the other hand, recent findings indicate that, under particular conditions, the metabolism of GSH by γ-GT can exert pro-oxidant effects [8, 122, 123].

GSH is synthesized inside cells and transported in the extracellular milieu through plasma-membrane transporters [130], with a concentration gradient (millimolar vs. micromolar). Extracellular metabolism of GSH by γ-GT, together with cell surface dipeptidases, promotes the release of constituent amino acids, the glutamic acid and the essential cysteine as well as their recovery by cells. Indeed, γ-GT overexpressing cells are able to utilize extracellular GSH as a source of cysteine more efficiently than normal cells [7], with a consequent selective growth advantage both at physiological and at limiting cysteine concentrations [131]. Thus, γ-GT can act as a source of essential amino acids, both for protein synthesis and for the maintenance of intracellular levels of GSH. Adequate levels of GSH play a key role in cellular resistance against several electrophilic/alkylating compounds. Indeed, γ-GT-overexpressing cells were shown to be more resistant to hydrogen peroxide, and chemotherapics such as cisplatin [132]. In melanoma cells, GSH depletion and γ-GT inhibition significantly increased cytotoxicity of oxidative stress conditions [133]. Interestingly, the same treatments were also shown to induce γ-GT expression, likely as a protective adaptation induced by oxidative stress itself. As such, γ-GT expression would perfectly fit with the so-called 'resistance phenotype', that is a common defense mechanism of transformed, preneoplastic cells, against injury by oxidants and xenobiotics [8]. However, conflicting results are reported on the supposed roles of GSH and γ-GT in protection against cell injury. Indeed, γ-GT could function in different ways, leading to extracellular detoxication of platinum-based drugs, but also to pro-oxidant effects catalyzed by metal ions in the extracellular space.

5.3 γ-GT and Cisplatin Resistance

Sulfur amino acids, in particular cysteine, and other small peptides containing cysteine, such as cysteinyl-glycine and GSH, are able to form adducts with cisplatin, and such complexes are poorly transported across plasma membrane. The effects are a decreased intracellular accumulation and a reduced toxicity of cisplatin towards treated cells [134]. Interestingly, cisplatin adducts with cysteinyl-glycine are formed ten times faster than those with GSH; such adducts are present in the extracellular medium of γ-GT overexpressing cells treated with cisplatin. The pKa of cysteinyl glycine thiol is significantly lower than that of GSH (6.4 vs. 8.6, respectively), thus causing its more rapid dissociation at physiological pH and its more efficient interaction with cisplatin. γ-GT transforms poorly reactive GSH into highly reactive cysteinyl-glycine, thus triggering the formation of cisplatin/thiol complexes in the extracellular space [135]. This results in lower cellular accumulation of cisplatin, reduced DNA platination, and reduced cytotoxicity [136]. Thus, the protective effect of γ-GT expression against cisplatin cytotoxicity seems to be dependent on the extracellular detoxication of cisplatin, rather than the higher intracellular GSH levels. However, it is likely that the rate of detoxification mediated by γ-GT may depend on the specific biological context, in which other resistance mechanisms can coexist. Indeed, cellular resistance to toxic agents acts as a multifactorial phenomenon, involving not only defense mechanisms, but also cellular response to genotoxic stress (DNA repair efficiency, DNA damage tolerance, stress response and susceptibility to apoptosis) [8].

Finally, dosage of cisplatin in vivo is limited by its nephrotoxicity, but the mechanism by which cisplatin kills proximal tubule cells remains unclear. However, both animal studies and clinical trials demonstrated that pre-treatment with exogenous GSH reduced nephrotoxicity induced by cisplatin without reducing its antitumor activity [129].

5.4 γ-GT and Tumor Progression

γ-GT can exert pro-oxidant effects at the membrane surface level and in the extracellular microenvironment [137]. This phenomenon was explained with the high reactivity of the γ-GT product, cysteinyl-glycine. Indeed, because of its lower pKa, the cysteinyl-glycine dissociates more quickly at physiological pH, reducing extracellular transition metal cations (in particular, Fe^{3+} and Cu^{2+}) more efficiently than GSH itself. Iron reduction by GSH, in fact, might be limited by the chelating properties of the alpha-carboxyl group of the glutamate residue, affecting steric and redox interactions of the cysteine thiol. The removal of glutamic acid by γ-GT causes a decrease of the cysteine thiol pKa and makes it free to interact with iron. The 'redox cycling' starting after iron reduction was shown to produce ROS and thiyl radicals, that is, reactive species promoting several intracellular and extra-cellular molecular effects. The damage induced by γ-GT could play an active role in the progress of preneoplastic foci to malignancy. Indeed, the pro-oxidant activity of γ-GT can promote the iron-dependent oxidative damage of DNA in transfected melanoma cells, thus contributing to genomic instability and increased mutation risk in cancer cells [138].

Moreover, γ-GT activity is able to promote the release of free iron from transferrin, thus promoting the uptake of iron by cancer cells [139]. This effect may play an additional role in supplying iron to malignant cells. Physiological levels of pro-oxidants can exert regulatory roles by acting on targets sensitive to the redox state of the cell [140]. A major role in such regulation is played by cysteine thiols, which can undergo different redox modifications [141, 142]. Such reactive cysteines play key role in proteins involved in crucial processes, such as cell proliferation, apoptosis, cell adhesion, and thus in progression of cancer and other diseases. It has been reported that γ-GT activity can promote the oxidation of thiol groups in cell surface proteins, through the involvement of hydrogen per-oxide and the formation of mixed disulfides ('protein S-thiolation') [141]. As hydrogen peroxide freely diffuses across the plasma membrane, pro-oxidant reactions dependent by γ-GT activity can also involve crucial intracellular targets, for example they can induce the binding of NF-κB and AP-1 to DNA [143, 144] and modulate the balance between protein kinase/phosphatase [145]. Indeed, it is well known that redox processes can play modulatory roles in the transduction of proliferative/apoptotic signals, due to interactions with growth factor receptors, protein kinases, and transcription factors.

The modulatory effects of pro-oxidant reactions could contribute to the resis-tance phenotype of γ-GT-expressing cancer cells, by regulating both signal transduction pathways involved in proliferation/apoptosis balance, as well as by inducing protective adaptations in the pool of intracellular antioxidants. For example, γ-GT/GSH-dependent pro-oxidant reactions were also shown to increase intracellular levels of vitamin C [146].

5.5 γ-GT as a Target for Anticancer Treatments

Because high intracellular levels of GSH are involved in the resistance phenotype of cancer cells, γ-GT has been considered a potential target of inhibition associated with chemotherapeutics. As described above, most of γ-GT inhibitors, such as Acivicin, Azaserine, boronate derivatives, L-glutamic acid derivatives; gamma-(monophenyl) phosphonoglutamate analogs [90] are toxic and cannot be used in humans. Recently, a novel class of uncompetitive inhibitors of γ-GT, less toxic than glutamine analogs, have been described [96]. However, the development of

γ-GT inhibitors with low toxicity remains an interesting perspective for pharmacological research, and could have a great impact on cancer therapy.

γ-GT expression and activity are also relevant in the pathophysiology of cellular processes involving NO and related compounds, such as GSNO. Indeed, treatments of human cancer cells with NO and NO mimetics can restore the sensitivity of resistant cell populations to the cytotoxic effects of chemotherapeutics [147]. Thus, NO acts as a 'chemosensitizing agent', likely by modulating processes associated with prevention or inhibition of cellular drug resistance mechanisms. Reactivation of NO signaling might in some way counteract the effects produced by hypoxia in solid tumors [148]. However, the mechanisms by which NO restores sensitivity to anticancer agents are not clearly understood. This critical action might be played by vascular changes (promotion of blood perfusion and tumor oxygenation), radical scavenging, down-regulation of the GSH detoxification, inhibition of key transcription factors, as well as inhibition of drug efflux transporters and DNA repair enzymes. Growth inhibition and chemosensitization to carboplatin treatments were observed after exposure of glioma cells to NONOates [149], while significant chemosensitization to cisplatin cytotoxicity was observed in cells transfected with inducible NO synthase (iNOS) gene [150]. S-Nitrosothiols, GSNO especially, are considered physiologic NO metabolites, able to transport NO in blood and tissues. γ-GT selectively metabolizes GSNO, thus promoting the release of NO [8, 151]. This fact may well be exploited in order to selectively target NO to γ-GT-expressing cancer cells (see medical applications).

5.6 γ-GT Complexes: Novel Biomarkers for Cancer and Other Pathologies

Serum γ-GT, considered as released exclusively from the liver, is widely used as a biomarker of liver dysfunction and excessive alcohol use [13]. On the other hand, several studies have revealed that γ-GT serum levels are positively associated with the risk of cardiovascular events [112], hypertension, type II diabetes and metabolic syndrome [113, 152], renal failure and cancer, even unrelated to hepatic involvement [153]. Therefore, diseased tissues other than the liver might contribute to serum γ-GT activity, thus explaining its broad predictive value. The release of γ-GT from cancer cells was described in several types of neoplasia, but the mechanisms by which cellular γ-GT is released in blood are still poorly characterized. Several papers investigated the possible specificity of serum γ-GT complex for certain tumors, in particular hepatocellular carcinoma, focusing on parameters such as γ-GT post-translational modifications and lipoprotein association. Specific γ-GT macroforms with clinical significance have been reported in patients with primary hepatocellular carcinoma, but the origin or structures of these complexes were not established [154, 155]. Recently, an in vitro study on melanoma and prostate cancer cells has shown the release of a γ-GT containing soluble complex with a MW >2000 kDa [156]. The component molecules of this complex are still to be identified. Variations in γ-GT glycosylation have been described when comparing the enzyme from malignant and normal tissues. These changes appear, however, to vary with the type of tumor [157], and the amount of tumor-derived γ-GT forms in serum may be affected by a rapid clearance rate [158]. Other studies are needed to better understand the properties of serum γ-GT fractions and the way they are released from cancer cells, in view of a clinical utilization of γ-GT as a biomarker of disease. In a retrospective study, total serum γ-GT was significantly increased in patients with metastatic renal cell carcinoma, but was normal in those with localized primary growths [159]. Similar results were obtained in other works, where both alkaline phosphatase and γ-GT activities were normal in a majority of patients with localized renal cell carcinoma, but increased in most of the patients with metastatic disease involving liver and/or bones. In both cases, γ-GT appeared to be a sensitive marker of metastatic renal cell carcinoma, even though not specific for the site of metastasis.

Significantly higher serum γ-GT levels were also found in hepatocellular carcinoma patients with poorly differentiated tumors. Nevertheless, serum γ-GT levels seem to be at least partly independent of γ-GT expression in tumors [123, 143], and their specificity as marker of cancer has been questioned [157]. On the other hand, epidemiologic studies have suggested that increased γ-GT levels can predict morbidity and mortality independently from liver diseases. Elevated γ-GT significantly increased cancer risk and, in site-specific cancer models, γ-GT was significantly associated with malignant neoplasms [158].

Finally, it has been recently found that soluble γ-GT may play a cytokine-like function. Indeed, the structure of γ-GT includes the chemokine-like CX_3C motif [20], which likely enables γ-GT to modulate bone resorption independently of its catalytic activity [160]. Moreover, urinary excretion of γ-GT is related to changes of biochemical markers of bone resorption [160]. Therefore, the overexpression and release of γ-GT by human tumors may have a role in establishment and development of bone metastasis.

6 γ-GTs: Biotechnological and Biomedical Applications

The biochemical ability of γ-GTs to cleave the unusual peptide bond of GSH and to transfer the γ-glutamyl moiety to some acceptors for producing γ-glutamyl compounds can be exploited in different ways for pharmaceutical and biotechnological applications. Since the γ-glutamyl linkage cannot be hydrolyzed by serum peptidases, the half lives of γ-glutamyl compounds are longer than those of non modified drugs [82]. Thus, the addition of a γ-glutamyl group at a specific site makes the drug inactive until the γ-glutamyl group is cleaved off by γ-GTs. Bacterial γ-GTs are a valuable group of enzymes which display a lot of advantageous industrial properties: their functional characteristics allow a great versatility of possible applications; they can be easily genetically manipulated; they are soluble (periplasmic or extracellular), non-glycosylated, and present a widely diversified substrate specificity, which can be used to employ different acceptors and synthesize various γ-glutamyl compounds. These proteins have been already successfully used as catalysts for the synthesis of pharmaceutically important peptides thanks to the transpeptidation reaction [82, 161]. Examples of the use of γ-GTs in pharmaceutical and biotechnological processes are reported below.

6.1 γ-GTs in the Production of SCV-07 and of Glutamyl Phenyl Hydrazine Analogs

γ-D-Glutamyl-L-tryptophan (SCV-07) is a potential drug for the treatment of tuberculosis, since it has been shown to possess abroad spectrum of immuno-stimulatory activities against murine tuberculosis [162]. This molecule is also a possible therapeutic for recurrent genital herpes simplex virus type 2 (Fig. 24) [163].

Because SCV-07 has several reactive groups and consists of both D- and L-amino acids, which are connected through a γ-glutamyl linkage, its chemical synthesis is complicated. An efficient enzymatic method to synthesize SCV-07 using not expensive reagents and employing bacterial γ-GTs has been developed [82, 161, 164]. The optimum reaction conditions are 50 mM D-glutamine, 50 mM L-tryptophan, and 0.2 U ml^{-1} γ-GT, pH 9–9.5, and incubation at 37 °C for 5 h. In these conditions, the reaction produces 33 mM γ-D-glutamyl-L-tryptophan, with a conversion rate of 66 %. The product can be purified by Dowex 1 × 8 column.

Another group of pharmacologically relevant molecules that can be synthesized using γ-GTs are the glutamyl phenyl hydrazine (GPH) derivatives. These

Fig. 24 Structure of SCV-07

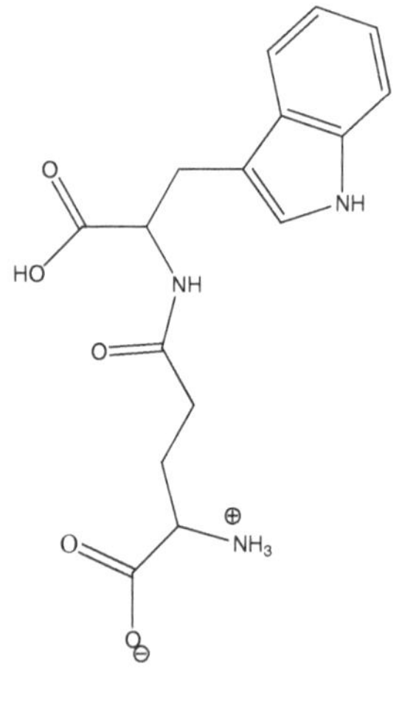

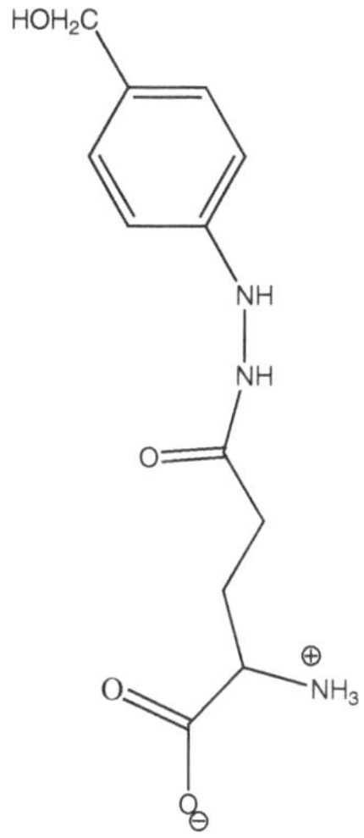

Fig. 25 Structure of
Agaritine

molecules, which exhibit antioxidant, immune-stimulatory, antiviral and anticancer effects, are generally obtained via chemical synthesis or by isolation from natural sources. Since γ-GTs are also involved in the synthesis of Agaritine (Fig. 25) occurring in *Agaricus bisporus* [165], it has been supposed that the GPH derivatives could be synthesized using phenyl hydrazine and L-glutamine through a transpeptidation reaction catalyzed by γ-GTs, and indeed this hypothesis has been proved for EcGT [166]. β-N-(γ-l(+)-glutamyl)-4-carboxyphenylhydrazine, a precursor of Agaritine, has been similarly prepared from L-glutamine and 4-hydrazinobenzoic acid using EcGT [167].

6.2 γ-GTs as Cephalosporin Acylases

Cephalosporins, the best-selling antibiotics worldwide, are derived from 7-aminocephalosporanic acid (7-ACA). Currently, in the pharmaceutical industry, 7-ACA is produced from cephalosporin C (CPC) by sequential application of D-amino acid oxidase and cephalosporin acylase (Fig. 26). In this two-step enzymatic process, D-amino acid oxidase catalyzes the oxidative deamination of CPC to 7-γ-(5-carboxy-5-oxypentanamide)-cephalosporanic acid, which then autoconvert to glutaryl-7-ACA (GL-7-ACA) [168]. Cephalosporin acylase (CA) then deacylates GL-7-ACA to 7-ACA (Fig. 26). A simpler and more environmentally sound conversion of CPC to 7-ACA by a single enzymatic reaction would be of great interest, but no enzyme catalyzes the direct bioconversion of CPC to 7-ACA at an acceptable rate. CAs are found in various *Pseudomonas* species and are classified into five classes on the basis of their sequence and

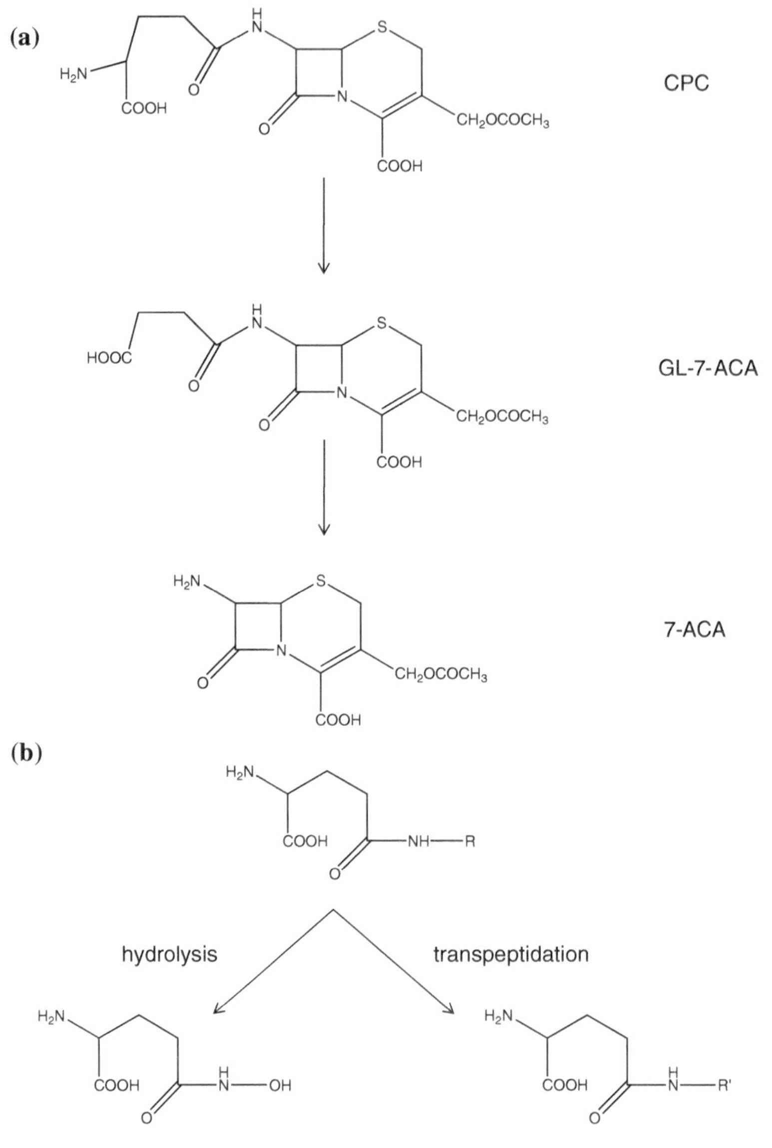

Fig. 26 a Enzymatic production of 7-ACA: oxidative deamination reaction catalyzed by D-amino acid oxidase and autoconversion to GL-7-ACA; deacylation reaction catalyzed by class IV CA; **b** Reactions catalyzed by γ-GTs: hydrolysis and transpeptidation

substrate specificity [164]. The primary structure of class IV CAs and γ-GTs is identical by about 30 % and some of these CAs show γ-GT activities. Thus, it has been speculated that Class IV CAs are γ-GTs with adventitious secondary CA activity [169].

Some γ-GTs present GL-7-ACA activity and other members of this enzyme family have been mutagenized to acquire this capability (Fig. 26). In particular, γ-GT from *Bacillus subtilis* possesses an inherent glutaryl-7-aminocephalosporanic acid acylase [40] activity with a k_{cat} value of 0.0485 s^{-1} and has been used for 7-ACA production. Site-directed and random mutagenesis has further increased its activity. In particular, it has been shown that k_{cat}/K_m value increases to 3.41 s^{-1} mM^{-1} and the k_{cat} value increases to 0.508 s^{-1} for the D445G and the E423Y/E442Q/D445N triple mutant, respectively. Consequently, the catalytic efficiency and the turnover rate have been improved up to about 1,000-fold and 10-fold, compared with the wildtype BsGT [170].

In EcGT, substitution of D433 abolishes transpeptidase activity, significantly reduces hydrolase activity, but imparts CA activity, which is absent in the wildtype enzyme [166]. In this frame, extremophilic γ-GTs represent natural mutants, lacking D433 and transpeptidase activity [46, 58].

6.3 γ-GTs as Glutaminases

The delicious taste of soy sauce, a traditional Japanese seasoning, mainly depends on the amount of L-glutamic acid, which is an important flavor-enhancing component. During soy sauce fermentation and in the manufacture of bread, proteins are digested into peptides via degradation by enzymes contained in the raw materials and among them, proteases. The peptides are then cleaved into amino-acids by specific peptidases. The glutamine liberated is hydrolyzed to glutamic acid by glutaminases, but when these enzymes are insufficient or inhibited, as often happens in the soy sauce fermentation, the glutamines are converted spontaneously to tasteless pyroglutamic acid.

γ-GTs can act as glutaminases, when they act as hydrolase, but lack transpeptidase activity, which could lead to the production of undesired by-products [33, 73–75]. In this respect, γ-GT resistant to high temperatures and high salt concentrations used during these processes are desirable [58, 76]. Minami, Suzuki, and Kumagai have produced and characterized a mutant of BsGT specialized in hydrolase activity, which could be used as a glutaminase in food industry [171]. GthGT also can function as glutaminase, since it is salt-tolerant [76], thermostable [100] and its transpeptidase activity is trascurable [46].

6.4 γ-GTs in the Production of Theanine

γ-L-glutamyl-ethylamide (5-N-ethyl-glutamine or theanine) is one of the major constituents of Japanese green and black teas (Fig. 27), known since 1949 [172] as

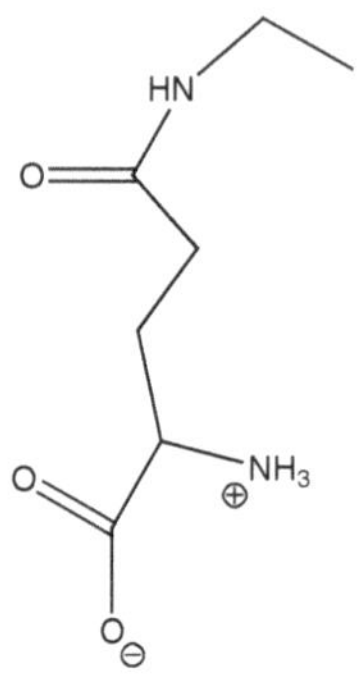

Fig. 27 Structure of
theanine

responsible of their taste. Recently, theanine [173] has drawn an increasing interest, especially for its potential medical applications, because it is able to decrease high blood pressure [174] and enhance the memory ability [175], to inhibit the excitation stimulated by caffeine [176] and improve attention [175], to enhance the effects of antitumor agents [177], to reduce stress [178], and to inhibit convulsive action [179]. Theanine is synthesized by theanine synthetase (EC 6.3.1.6) in plants. Since it is expensive and not readily available, several methods have been developed for its production [180]. One of the most effective methods is theanine isolation from dry tea or fresh tea plant leaves, which have high theanine content. However, this procedure is difficult, time-consuming, and expensive. Other methods such as tea callus cultivation [181] and fermentation of the mushroom *Xerocomus badius* [182] have also been reported, but the production efficiency of these procedures is low. Chemical synthesis of theanine has been performed for the first time in 1942 [183] with a yield of 90 g kg^{-1} by treating pyrrolidone-5-carboxylic acid with aqueous ethylamine for 20 days at 37 °C. Since then, many other chemical approaches have been carried out, but all have limitations including low yield and purity, and requirement for protection and unblocking of reactive groups [184, 185]. Consequently, there is a demand for other methods of production of theanine. Enzymatic theanine synthesis using γ-GT from *E. coli* and glutamine as substrate have been reported [25, 42, 186]. Similar experiments have proved that theanine can be synthesized also using γ-GT from *P. nitroreducens* [69] or γ-GT from *Bacillus subtilis* SK11.004 [70]. An alternative approach uses glutamic acid γ-methyl ester and immobilized *E. coli* cells with γ-GT activity as catalyst. The results show that activity is about 1.2-fold higher with glutamic acid γ-methyl ester than that with glutamine as substrate. Reaction conditions have been optimized at pH 10 and 50 °C. Under these conditions, the immobilized cells are continuously used 10 times, yielding an average glutamic acid γ-methyl ester to theanine conversion rate of 69.3 %.

6.5 γ-GT to Release Drugs

γ-GT has been proposed as a tool to improve the therapeutic efficacy of selected drugs. Thanks to its differential expression in several tumors, this enzyme can be exploited to design pro-drugs that can selectively enter to cells of γ-GT-positive tumors.

As above mentioned, GSNO can function as stable transporter of NO in blood and tissues. Since γ-GT selectively metabolizes GSNO, it can promote the NO release [8, 151]. Indeed, GSNO is a suitable pro-drug to selectively target NO to γ-GT-expressing tumors [151], in fact, the K_m of γ-GT with respect to GSNO is approximately 0.4 mM, comparable with K_m value for GSH.

Since γ-GT is primarily expressed in the kidney (expression levels are 5-/10-fold higher than in the liver and pancreas), the same approach has been used to trigger reno-selective release of drugs (Fig. 28) [187]. NO releasing molecules with a γ-glutamyl group activated by γ-GT represent potential drugs for the treatment of acute renal injury/failure, a common complication in intensive care unit [188]. γ-GT catalyzes the conversion of GSNO to S-nitrocysteinglycine (GCSNO), which releases NO in the absence of a chelating agent. A similar strategy has been recently described for other reno-selective NO-donors [189]. Similarly, pro-drugs activated by γ-GT to release γ-L-glutamyl-dihydroxyphenylalanine (L-DOPA) in the kidneys have been reported [190]. L-DOPA could be used to increase the concentration of dopamine in the brain, and for this reason it is a promising pro-drug for Parkinson's disease [191].

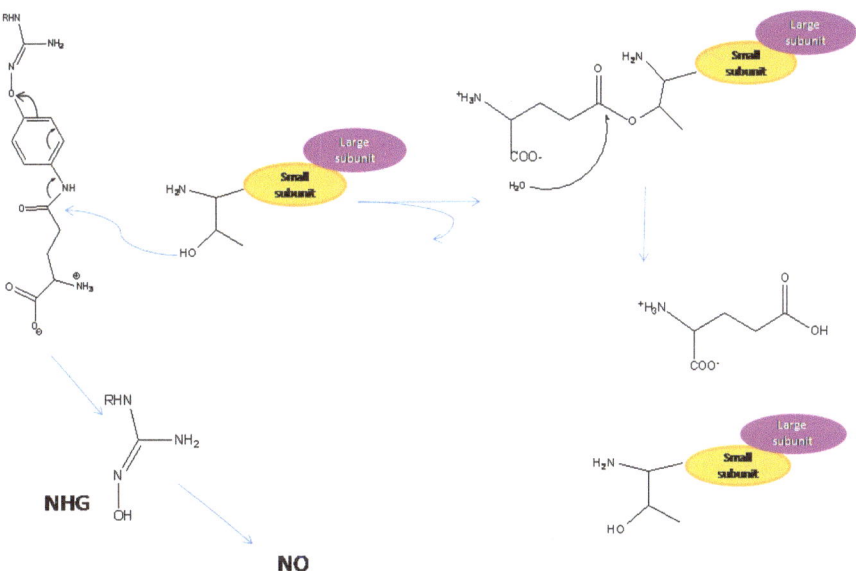

Fig. 28 Generation of NO by hydrolysis of γ-glutamyl bonds of N-hydroxyguanidine (NHG) derivatives by γ-GT. NHG is used as NO-releasing molecule

7 Conclusions

Structural and functional studies on γ-GTs have provided important information about the autocatalysis and enzymatic reaction mechanisms of these enzymes. However, despite considerable progresses have been made in the understanding of their structure–function relationships, there are still many unanswered questions. Indeed, most of the literature on γ-GTs was focused on understanding how the enzymes were autoprocessed and on establishing details of their catalytic mechanism, including the identification of residues that form the γ-glutamyl binding site. However, further kinetic and structural studies on these enzymes are needed to determine details of the mechanisms and relative importance of the transpeptidation and hydrolysis reactions. In this respect, it is important to remark that mechanism of transpeptidation still remains unclear and that residues involved in recognition of acceptor and in formation of the acceptor binding site have not been elucidated. Unraveling the reaction mechanisms of γ-GTs remains one of the major challenges for the rationale design of specific and potent drugs/inhibitors for this class of enzymes. Moreover, the physiological role of the hydrolysis reaction seems to be universally accepted, while that of transpeptidation remains controversial. In mammals, γ-GT catalyze the γ-glutamylation of the amino acids produced upon the hydrolysis of extracellular GSH, and this transfer reaction seems to be a strategy to favor their re-uptake in the cell. In unicellular organisms, the transpeptidase reaction is poor or absent and could play a marginal role. These speculations need further studies to be confirmed.

Virtually nothing is known about the structures of the mammalian membrane-embedded γ-GTs. There are no studies that establish if oligomerization of the precursor and/or hetero-oligomerization of the mature forms are important in controlling enzyme activity and cellular localization. Moreover, the possible presence of sites for 'allosteric' regulation has been not still determined. The interplay between the small and the large subunits needs also to be addressed and further investigations on the possibility that these enzymes could be also involved in different metabolic pathways are needed.

Despite recent advancements in predicting the catalytic function of γ-GT mutants, there are big limitations in gain of function studies, probably because of our lack of understanding the molecular mechanisms that define protein behavior in space and time, that is, as flexible and dynamic macromolecules. In this respect, it should be noted that dynamic behavior of γ-GTs has received scarce attention from investigators. New experiments should be performed to elucidate the dynamics of γ-GTs and then assess the contributions of conformational changes and concerted residue motions to enzyme catalysis.

Finally, thermodynamic data on the folding/unfolding pathways of the precursor, as well as on those of the mature protein, could be interesting, since they will provide extremely useful information on the energetic forces that drive the folding and the assembly of the subunits upon the autoprocessing.

References

1. Foyer CH, Noctor G (2005) Redox homeostasis and antioxidant signaling: a metabolic interface between stress perception and physiological responses. Plant Cell 17(7):1866–1875
2. Meister A, Anderson ME (1983) Glutathione. Annu Rev Biochem 52:711–760
3. Folk JE (1969) Mechanism of action of guinea pig liver transglutaminase. VI. Order of substrate addition. J Biol Chem 244(13):3707–3713
4. Meister A (1973) On the enzymology of amino acid transport. Science 180(4081):33–39
5. Pompella A, De Tata V, Paolicchi A, Zunino F (2006) Expression of gamma-glutamyltransferase in cancer cells and its significance in drug resistance. Biochem Pharmacol 71(3):231–238
6. Meister A, Tate SS, Griffith OW (1981) Gamma-glutamyl transpeptidase. Methods Enzymol 77:237–253
7. Lieberman MW, Wiseman AL, Shi ZZ, Carter BZ, Barrios R, Ou CN, Chevez-Barrios P, Wang Y, Habib GM, Goodman JC, Huang SL, Lebovitz RM, Matzuk MM (1996) Growth retardation and cysteine deficiency in gamma-glutamyl transpeptidase-deficient mice. Proc Natl Acad Sci U S A 93(15):7923–7926
8. Pompella A, Corti A, Paolicchi A, Giommarelli C, Zunino F (2007) Gamma-glutamyltransferase, redox regulation and cancer drug resistance. Curr Opin Pharmacol 7(4):360–366
9. Iannaccone PM, Koizumi J (1983) Pattern and rate of disappearance of gamma-glutamyl transpeptidase activity in fetal and neonatal rat liver. J Histochem Cytochem 31(11):1312–1316
10. Cotariu D, Barr-Nea L, Papo N, Zaidman JL (1988) Induction of gamma-glutamyl transferase by dexamethasone in cultured fetal rat hepatocytes. Enzyme 40(4):212–216
11. Tongiani R, Paolicchi A (1989) Gamma-glutamyltransferase induction by glucocorticoids in rat liver: age-dependence, time-dependence, dose-dependence, and intralobular distribution. Acta Histochem 86(1):51–61
12. Brouillet A, Holic N, Chobert MN, Laperche Y (1998) The gamma-glutamyl transpeptidase gene is transcribed from a different promoter in rat hepatocytes and biliary cells. Am J Pathol 152(4):1039–1048
13. Whitfield JB (2001) Gamma glutamyl transferase. Crit Rev Clin Lab Sci 38(4):263–355
14. Courtay C, Heisterkamp N, Siest G, Groffen J (1994) Expression of multiple gamma-glutamyltransferase genes in man. Biochem J 297(Pt 3):503–508
15. Figlewicz DA, Delattre O, Guellaen G, Krizus A, Thomas G, Zucman J, Rouleau GA (1993) Mapping of human gamma-glutamyl transpeptidase genes on chromosome 22 and other human autosomes. Genomics 17(2):299–305
16. Visvikis A, Pawlak A, Accaoui MJ, Ichino K, Leh H, Guellaen G, Wellman M (2001) Structure of the 5′ sequences of the human gamma-glutamyltransferase gene. Eur J Biochem 268(2):317–325
17. Diederich M, Wellman M, Visvikis A, Puga A, Siest G (1993) The 5′ untranslated region of the human gamma-glutamyl transferase mRNA contains a tissue-specific active translational enhancer. FEBS Lett 332(1–2):88–92
18. Farber E (1984) Cellular biochemistry of the stepwise development of cancer with chemicals: G. H. A. Clowes memorial lecture. Cancer Res 44 (12 Pt 1):5463–5474
19. Farber E (1984) The multistep nature of cancer development. Cancer Res 44(10):4217–4223
20. Niida S, Kawahara M, Ishizuka Y, Ikeda Y, Kondo T, Hibi T, Suzuki Y, Ikeda K, Taniguchi N (2004) Gamma-glutamyltranspeptidase stimulates receptor activator of nuclear factor-kappaB ligand expression independent of its enzymatic activity and serves as a pathological bone-resorbing factor. J Biol Chem 279(7):5752–5756
21. Hanes CS, Hird FJ (1950) Synthesis of peptides in enzymic reactions involving glutathione. Nature 166(4216):288–292

22. Okada T, Suzuki H, Wada K, Kumagai H, Fukuyama K (2006) Crystal structures of gamma-glutamyltranspeptidase from *Escherichia coli*, a key enzyme in glutathione metabolism, and its reaction intermediate. Proc Natl Acad Sci U S A 103(17):6471–6476

23. Boanca G, Sand A, Okada T, Suzuki H, Kumagai H, Fukuyama K, Barycki JJ (2007) Autoprocessing of *Helicobacter pylori* gamma-glutamyltranspeptidase leads to the formation of a threonine-threonine catalytic dyad. J Biol Chem 282(1):534–541

24. Nakayama R, Kumagai H, Tochikura T (1984) Purification and properties of gamma-glutamyltranspeptidase from *Proteus mirabilis*. J Bacteriol 160(1):341–346

25. Suzuki H, Kumagai H, Tochikura T (1986) Gamma-glutamyltranspeptidase from *Escherichia coli* K-12: purification and properties. J Bacteriol 168(3):1325–1331

26. Ogawa Y, Hosoyama H, Hamano M, Motai H (1991) Purification and properties of gamma-glutamyltranspeptidase from *Bacillus subtilis* (natto). Agric Biol Chem 55(12):2971–2977

27. Chevalier C, Thiberge JM, Ferrero RL, Labigne A (1999) Essential role of *Helicobacter pylori* gamma-glutamyltranspeptidase for the colonization of the gastric mucosa of mice. Mol Microbiol 31(5):1359–1372

28. Martin MN, Slovin JP (2000) Purified gamma-glutamyl transpeptidases from tomato exhibit high affinity for glutathione and glutathione S-conjugates. Plant Physiol 122(4):1417–1426

29. Lancaster JE, Shaw ML (1994) Characterization of purified gamma-glutamyl transpeptidase in onions: evidence for in vivo role as peptidase. Phytochemistry 36:1351–1358

30. West MB, Hanigan MH (2010) γ-Glutamyl transpeptidase is a heavily N-glycosylated heterodimer in HepG2 cells. Arch Biochem Biophys 504(2):177–181

31. Tate SS, Meister A (1985) Gamma-glutamyl transpeptidase from kidney. Methods Enzymol 113:400–419

32. Ikeda Y, Fujii J, Taniguchi N, Meister A (1995) Expression of an active glycosylated human gamma-glutamyl transpeptidase mutant that lacks a membrane anchor domain. Proc Natl Acad Sci U S A 92(1):126–130

33. Minami H, Suzuki H, Kumagai H (2003) A mutant *Bacillus subtilis* gamma-glutamyltranspeptidase specialized in hydrolysis activity. FEMS Microbiol Lett 224(2):169–173

34. West MB, Wickham S, Quinalty LM, Pavlovicz RE, Li C, Hanigan MH (2011) Autocatalytic cleavage of human gamma-glutamyl transpeptidase is highly dependent on N-glycosylation at asparagine 95. J Biol Chem 286(33):28876–28888

35. Kinlough CL, Poland PA, Bruns JB, Hughey RP (2005) Gamma-glutamyltranspeptidase: disulfide bridges, propeptide cleavage, and activation in the endoplasmic reticulum. Methods Enzymol 401:426–449

36. Wu R, Richter S, Zhang RG, Anderson VJ, Missiakas D, Joachimiak A (2009) Crystal structure of *Bacillus anthracis* transpeptidase enzyme CapD. J Biol Chem 284(36):24406–24414

37. Brannigan JA, Dodson G, Duggleby HJ, Moody PC, Smith JL, Tomchick DR, Murzin AG (1995) A protein catalytic framework with an N-terminal nucleophile is capable of self-activation. Nature 378(6555):416–419

38. Oinonen C, Rouvinen J (2000) Structural comparison of Ntn-hydrolases. Protein Sci 9(12):2329–2337

39. Okada T, Suzuki H, Wada K, Kumagai H, Fukuyama K (2007) Crystal structure of the gamma-glutamyltranspeptidase precursor protein from *Escherichia coli*. Structural changes upon autocatalytic processing and implications for the maturation mechanism. J Biol Chem 282(4):2433–2439

40. Wada K, Irie M, Suzuki H, Fukuyama K (2010) Crystal structure of the halotolerant gamma-glutamyltranspeptidase from *Bacillus subtilis* in complex with glutamate reveals a unique architecture of the solvent-exposed catalytic pocket. FEBS J 277(4):1000–1009

41. Williams K, Cullati S, Sand A, Biterova EI, Barycki JJ (2009) Crystal structure of acivicin-inhibited gamma-glutamyltranspeptidase reveals critical roles for its C-terminus in autoprocessing and catalysis. Biochemistry 48(11):2459–2467

42. Suzuki H, Kajimoto Y, Kumagai H (2002) Improvement of the bitter taste of amino acids through the transpeptidation reaction of bacterial gamma-glutamyltranspeptidase. J Agric Food Chem 50(2):313–318

43. Chang HP, Liang WC, Lyu RC, Chi MC, Wang TF, Su KL, Hung HC, Lin LL (2010) Effects of C-terminal truncation on autocatalytic processing of *Bacillus licheniformis* gamma-glutamyl transpeptidase. Biochemistry (Mosc) 75(7):919–929

44. Lin LL, Chou PR, Hua YW, Hsu WH (2006) Overexpression, one-step purification, and biochemical characterization of a recombinant gamma-glutamyltranspeptidase from *Bacillus licheniformis*. Appl Microbiol Biotechnol 73(1):103–112

45. Angele C, Oster T, Visvikis A, Michels JM, Wellman M, Siest G (1991) Different constructs for the expression of mammalian gamma-glutamyltransferase cDNAs in *Escherichia coli* and in *Saccharomyces cerevisiae*. Clin Chem 37(5):662–666

46. Castellano I, Merlino A, Rossi M, La Cara F (2010) Biochemical and structural properties of gamma-glutamyl transpeptidase from *Geobacillus thermodenitrificans*: an enzyme specialized in hydrolase activity. Biochimie 92(5):464–474

47. Boanca G, Sand A, Barycki JJ (2006) Uncoupling the enzymatic and autoprocessing activities of *Helicobacter pylori* gamma-glutamyltranspeptidase. J Biol Chem 281(28):19029–19037

48. Suzuki H, Kumagai H (2002) Autocatalytic processing of gamma-glutamyltranspeptidase. J Biol Chem 277:43536

49. Hashimoto W, Suzuki H, Nohara S, Tachi H, Yamamoto K, Kumagai H (1995) Subunit association of gamma-glutamyltranspeptidase of *Escherichia coli* K-12. J Biochem 118(6):1216–1223

50. Hashimoto W, Suzuki H, Yamamoto K, Kumagai H (1995) Effect of site-directed mutations on processing and activity of gamma-glutamyltranspeptidase of *Escherichia coli* K-12. J Biochem 118(1):75–80

51. Meng-Chun Chi MC, Chen YY, Lo HF, Lin LL (2012) Experimental evidence for the involvement of amino acid residue Glu398 in the autocatalytic processing of *Bacillus licheniformis* γ-glutamyltranspeptidase. FEBS Open Bio 2:298–304

52. Hashimoto W, Suzuki H, Nohara S, Kumagai H (1992) *Escherichia coli* gamma-glutamyltranspeptidase mutants deficient in processing to subunits. Biochem Biophys Res Commun 189(1):173–178

53. Ikeda Y, Fujii J, Taniguchi N (1996) Effects of substitutions of the conserved histidine residues in human gamma-glutamyl transpeptidase. J Biochem 119(6):1166–1170

54. Keillor JW, Castonguay R, Lherbet C (2005) Gamma-glutamyl transpeptidase substrate specificity and catalytic mechanism. Methods Enzymol 401:449–467

55. Lyu RC, Hu HY, Kuo LY, Lo HF, Ong PL, Chang HP, Lin LL (2009) Role of the conserved Thr399 and Thr417 residues of *Bacillus licheniformis* gamma-glutamyltranspeptidase as evaluated by mutational analysis. Curr Microbiol 59(2):101–106

56. Ikeda Y, Fujii J, Anderson ME, Taniguchi N, Meister A (1995) Involvement of Ser-451 and Ser-452 in the catalysis of human gamma-glutamyl transpeptidase. J Biol Chem 270(38):22223–22228

57. Ikeda Y, Fujii J, Taniguchi N, Meister A (1995) Human gamma-glutamyl transpeptidase mutants involving conserved aspartate residues and the unique cysteine residue of the light subunit. J Biol Chem 270(21):12471–12475

58. Castellano I, Di Salle A, Merlino A, Rossi M, La Cara F (2011) Gene cloning and protein expression of gamma-glutamyltranspeptidases from *Thermus thermophilus* and *Deinococcus radiodurans*: comparison of molecular and structural properties with mesophilic counterparts. Extremophiles 15(2):259–270

59. Hu X, Legler PM, Khavrutskii I, Scorpio A, Compton JR, Robertson KL, Friedlander AM, Wallqvist A (2012) Probing the donor and acceptor substrate specificity of the gamma-glutamyl transpeptidase. Biochemistry 51(6):1199–1212

60. Taniguchi N, Ikeda Y (1998) Gamma glutamyl transpeptidase: catalytic mechanism and gene expression. Adv Enzymol Relat Areas Mol Biol 72:239–278

61. Thompson GA, Meister A (1977) Interrelationships between the binding sites for amino acids, dipeptides, and gamma-glutamyl donors in gamma-glutamyl transpeptidase. J Biol Chem 252(19):6792–6798

62. Rajput R, Verma VV, Chaudhary V, Gupta R (2013) A hydrolytic gamma-glutamyl transpeptidase from thermo-acidophilic archaeon *Picrophilus torridus*: binding pocket mutagenesis and transpeptidation. Extremophiles 17(1):29–41

63. Lherbet C, Gravel C, Keillor JW (2004) Synthesis of S-alkyl L-homocysteine analogues of glutathione and their kinetic studies with gamma-glutamyl transpeptidase. Bioorg Med Chem Lett 14(13):3451–3455

64. Keillor JW, Menard A, Castonguay R et al (2004) Pre-steady-state kinetic studies of rat kidney gamma-glutamyl transpeptidase confirm its ping-pong mechanism. J Phys Org Chem 17:529–536

65. Orlowski M, Meister A (1963) Gamma-glutamyl-*p*-nitroanilide: a new convenient substrate for determination and study of L- and D-gamma-glutamyltranspeptidase activities. Biochim Biophys Acta 73:679–681

66. Wickham S, West MB, Cook PF, Hanigan MH (2011) Gamma-glutamyl compounds: substrate specificity of gamma-glutamyl transpeptidase enzymes. Anal Biochem 414(2):208–214

67. Beutler HO (1990) L-glutamate, colorimetric method with glutamate dehydrogenase and diaphorase. Bergmeyer HU (ed) Methods of enzymatic analysis. VCH, Cambridge, pp 369–376

68. Storozhenko S, Belles-Boix E, Babiychuk E, Herouart D, Davey MW, Slooten L, Van Montagu M, Inze D, Kushnir S (2002) Gamma-glutamyl transpeptidase in transgenic tobacco plants. Cellular localization, processing, and biochemical properties. Plant Physiol 128(3):1109–1119

69. Imaoka M, Yano S, Okumura M, Hibi T, Wakayama M (2010) Molecular cloning and characterization of gamma-glutamyltranspeptidase from *Pseudomonas nitroreducens* IFO12694. Biosci Biotechnol Biochem 74(9):1936–1939

70. Shuai Y, Zhang T, Mu W, Jiang B (2011) Purification and characterization of gamma-glutamyltranspeptidase from *Bacillus subtilis* SK11.004. J Agric Food Chem 59:6233–6238

71. Abe K, Ito Y, Ohmachi T, Asada Y (1997) Purification and properties of two isozymes of gamma-glutamyltranspeptidase from *Bacillus subtilis* TAM-4. Biosci Biotechnol Biochem 61(10):1621–1625

72. Murty NA, Tiwary E, Sharma R, Nair N, Gupta R (2011) Gamma-glutamyl transpeptidase from *Bacillus pumilus* KS 12: decoupling autoprocessing from catalysis and molecular characterization of N-terminal region. Enzyme Microb Technol 50(3):159–164

73. Tiwary E, Gupta R (2010) Improved catalytic efficiency of a monomeric gamma-glutamyl transpeptidase from *Bacillus licheniformis* in presence of subtilisin. Biotechnol Lett 32(8):1137–1141

74. Tiwary E, Gupta R (2010) Subtilisin-gamma-glutamyl transpeptidase: a novel combination as ungual enhancer for prospective topical application. J Pharm Sci 99(12):4866–4873

75. Yang JC, Liang WC, Chen YY, Chi MC, Lo HF, Chen HL, Lin LL (2011) Biophysical characterization of *Bacillus licheniformis* and *Escherichia coli* gamma-glutamyltranspeptidases: a comparative analysis. Int J Biol Macromol 48(3):414–422

76. Pica A, Russo Krauss I, Castellano I, La Cara F, Graziano G, Sica F, Merlino A (2013) Effect of NaCl on the conformational stability of the thermophilic gamma-glutamyltranspeptidase from *Geobacillus thermodenitrificans*: implication for globular protein halotolerance. Biochim Biophys Acta 1834(1):149–157

77. Madern D, Ebel C, Zaccai G (2000) Halophilic adaptation of enzymes. Extremophiles 4(2):91–98

78. Madern D, Ebel C, Mevarech M, Richard SB, Pfister C, Zaccai G (2000) Insights into the molecular relationships between malate and lactate dehydrogenases: structural and biochemical properties of monomeric and dimeric intermediates of a mutant of tetrameric

L-[LDH-like] malate dehydrogenase from the halophilic archaeon *Haloarcula marismortui.* Biochemistry 39(5):1001–1010

79. Tate SS, Meister A (1974) Stimulation of the hydrolytic activity and decrease of the transpeptidase activity of gamma-glutamyl transpeptidase by maleate; identity of a rat kidney maleate-stimulated glutaminase and gamma-glutamyl transpeptidase. Proc Natl Acad Sci U S A 71(9):3329–3333

80. Thompson GA, Meister A (1980) Modulation of gamma-glutamyl transpeptidase activities by hippurate and related compounds. J Biol Chem 255(5):2109–2113

81. Gardell SJ, Tate SS (1983) Effects of bile acids and their glycine conjugates on gamma-glutamyl transpeptidase. J Biol Chem 258(10):6198–6201

82. Suzuki H, Yamada C, Kato K (2007) Gamma-glutamyl compounds and their enzymatic production using bacterial gamma-glutamyltransferase. Amino Acids 32(3):333–340

83. Segel GB, Woodlock TJ, Murant FG, Lichtman MA (1989) Photoinhibition of 2-amino-2-carboxybicyclo [2,2,1]heptane transport by O-diazoacetyl-L-serine. An initial step in identifying the L-system amino acid transporter. J Biol Chem 264(28):16399–16402

84. Hanigan MH, Gallagher BC, Taylor PT Jr, Large MK (1994) Inhibition of gamma-glutamyl transpeptidase activity by acivicin in vivo protects the kidney from cisplatin-induced toxicity. Cancer Res 54(22):5925–5929

85. Cutrín JC, Zingaro B, Camandola S, Boveris A, Pompella A, Poli G (2000) Contribution of γ-glutamyl transpeptidase to oxidative damage of ischemic rat kidney. Kidney Int 57:526–533

86. Chittur SV, Klem TJ, Shafer CM, Davisson VJ (2001) Mechanism for acivicin inactivation of triad glutamine amidotransferases. Biochemistry 40:876–887

87. Earhart RH, Neil GL (1985) Acivicin in 1985. Adv Enzyme Regul 24:179–205

88. Antczak C, Karp DR, London RE, Bauvois B (2001) Reanalysis of the involvement of gamma-glutamyl transpeptidase in the cell activation process. FEBS Lett 508(2):226–230

89. King JB, West MB, Cook PF, Hanigan MH (2009) A novel, species-specific class of uncompetitive inhibitors of gamma-glutamyl transpeptidase. J Biol Chem 284:9059–9065

90. Han L, Hiratake J, Kamiyama A, Sakata K (2007) Design, synthesis, and evaluation of gamma-phosphono diester analogues of glutamate as highly potent inhibitors and active site probes of gamma-glutamyl transpeptidase. Biochemistry 46(5):1432–1447

91. Yamamoto S, Watanabe B, Hiratake J, Tanaka R, Ohkita M, Matsumura Y (2011) Preventive effect of GGsTop, a novel and selective gamma-glutamyl transpeptidase inhibitor, on ischemia/reperfusion-induced renal injury in rats. J Pharmacol Exp Ther 339(3):945–951

92. Stein RL, DeCicco C, Nelson D, Thomas B (2001) Slow-binding inhibition of gamma-glutamyl transpeptidase by gamma-boroGlu. Biochemistry 40(19):5804–5811

93. Suenaga H, Nakashima K, Mikami M, Yamamoto H, James TD, Sandanayake KRAS, Shinkai S (1996) Screening of arylboronic acids to search for a strong inhibitor for γ-glutamyl transpeptidase (γ-GTP). Recl Trav Chim Pays-Bas 115:44–48

94. Tate SS, Meister A (1978) Serine-borate complex as a transition-state inhibitor of gamma-glutamyl transpeptidase. Proc Natl Acad Sci U S A 75(10):4806–4809

95. Wickham S, Regan N, West MB, Thai J, Cook PF, Terzyan SS, Li PK, Hanigan MH (2013) Inhibition of human gamma-glutamyl transpeptidase: development of more potent, physiologically relevant, uncompetitive inhibitors. Biochem J 450(3):547–557

96. King JB, West MB, Cook PF, Hanigan MH (2009) A novel, species-specific class of uncompetitive inhibitors of gamma-glutamyl transpeptidase. J Biol Chem 284(14):9059–9065

97. Wada K, Hiratake J, Irie M, Okada T, Yamada C, Kumagai H, Suzuki H, Fukuyama K (2008) Crystal structures of *Escherichia coli* gamma-glutamyltransferase in complex with azaserine and acivicin: novel mechanistic implication for inhibition by glutamine antagonists. J Mol Biol 380(2):361–372

98. Tate SS, Meister A (1977) Affinity labeling of gamma-glutamyl transpeptidase and location of the gamma-glutamyl binding site on the light subunit. Proc Natl Acad Sci U S A 74(3):931–935

99. Hsu WH, Ong PL, Chen SC, Lin LL (2009) Contribution of Ser463 residue to the enzymatic and autoprocessing activities of *Escherichia coli* gamma-glutamyltranspeptidase. Indian J Biochem Biophys 46(4):281–288

100. Pica A, Russo Krauss I, Castellano I, Rossi M, La Cara F, Graziano G, Sica F, Merlino A (2012) Exploring the unfolding mechanism of γ-glutamyltranspeptidases: the case of the thermophilic enzyme from *Geobacillus thermodenitrificans*. Biochim Biophys Acta Proteins Proteomics 1824(4):571–577

101. Van Ho T, Kamei K, Wada K, Fukuyama K, Suzuki H (2013) Thermal denaturation and renaturation of gamma-glutamyltranspeptidase of *Escherichia coli*. Biosci Biotechnol Biochem 77(2):409–412

102. Castellano I, Merlino A (2012) Gamma-glutamyltranspeptidases: sequence, structure, biochemical properties, and biotechnological applications. Cell Mol Life Sci 69(20):3381–3394

103. Betro MG, Oon RC, Edwards JB (1973) Gamma-glutamyl transpeptidase in diseases of the liver and bone. Am J Clin Pathol 60(5):672–678

104. Betro MG, Oon RC, Edwards JB (1973) Gamma-glutamyl transpeptidase and other liver function tests in myocardial infarction and heart failure. Am J Clin Pathol 60(5):679–683

105. Lee DH, Ha MH, Kim JH, Christiaani DC, Gross M, Steffes M, Blomkoff R, Jacobs DR Jr (2003) Gamma-glutamyltransferase and diabetes—a 4 year follow-up study. Diabetologia 46:359–364

106. Hashimoto Y, Futamura A, Nakarai H, Nakahara K (2001) Relationship between response of gamma-glutamyl transpeptidase to alcohol drinking and risk factors for coronary heart disease. Atherosclerosis 158(2):465–470

107. Lee DH, Jacobs DR Jr, Gross M, Kiefe CI, Roseman J, Lewis CE, Steffes M (2003) Gamma-glutamyltransferase is a predictor of incident diabetes and hypertension: the coronary artery risk development in young adults (CARDIA) study. Clin Chem 49(8):1358–1366

108. Liu CF, Gu YT, Wang HY, Fang NY Gamma-glutamyltransferase level and risk of hypertension: a systematic review and meta-analysis. PLoS ONE 7(11):e48878

109. Franzini M, Corti A, Martinelli B, Del CA, Emdin M et al (2009) Gamma-glutamyltransferase activity in human atherosclerotic plaques—biochemical similarities with the circulating enzyme. Atherosclerosis 202:119–127

110. Lee DH, Blomhoff R, Jacobs DR (2004) Is serum gamma glutamyltransferase a marker of oxidative stress? Free Radical Res 38:535–539

111. Turgut O, Yilmaz MB, Yalta K, Tandogan I (2009) Gamma-glutamyltransferase as a useful predictor for cardiovascular risk: clinical and epidemiological perspectives. Atherosclerosis 202(2):348–349

112. Emdin M, Passino C, Pompella A, Paolicchi A (2006) Gamma-glutamyltransferase as a cardiovascular risk factor. Eur Heart J 27(18):2145–2146

113. Lee DS, Evans JC, Robins SJ, Wilson PW, Albano I, Fox CS, Wang TJ, Benjamin EJ, D'Agostino RB, Vasan RS (2007) Gamma glutamyl transferase and metabolic syndrome, cardiovascular disease, and mortality risk: the Framingham heart study. Arterioscler Thromb Vasc Biol 27(1):127–133

114. Marchesini G, Avagnina S, Barantani EG, Ciccarone AM, Corica F, Dall'Aglio E, Dalle Grave R, Morpurgo PS, Tomasi F, Vitacolonna E (2005) Aminotransferase and gamma-glutamyltranspeptidase levels in obesity are associated with insulin resistance and the metabolic syndrome. J Endocrinol Invest 28:333–339

115. Lum G, Gambino SR (1972) Serum gamma-glutamyltranspeptidase activity as an indicator of disease of liver, pancreas, or bone. Clin Chem 18:358–362

116. Cabrera-Abreu JC, Green A (2002) Gamma-glutamyltransferase: value of its measurement in paediatrics. Ann Clin Biochem 39(Pt 1):22–25

117. Owen AD, Schapira AH, Jenner P, Marsden CD (1996) Oxidative stress and Parkinson's disease. Ann N Y Acad Sci 786:217–223
118. Paolicchi A, Minotti G, Tonarelli P, Tongiani R, De Cesare D, Mezzetti A, Dominici S, Comporti M, Pompella A (1999) Gamma-glutamyltranspeptidase-dependent iron reduction and LDL oxidation—a potential mechanism in atherosclerosis. J Investig Med 47:151–160
119. Zhang H, Forman HJ (2009) Redox regulation of gamma-glutamyltranspeptidase. Am J Respir Cell Mol Biol 41(5):509–515
120. Jean JC, Harding CO, Oakes SM, Yu Q, Held PK, Joyce-Brady M (1999) Gamma-glutamyl transferase (GGT) deficiency in the GGTenu1 mouse results from a single point mutation that leads to a stop codon in the first coding exon of GGT mRNA. Mutagenesis 14(1):31–36
121. Hanigan MH, Frierson HF Jr, Swanson PE, De Young BR (1999) Altered expression of gamma-glutamyl transpeptidase in human tumors. Hum Pathol 30(3):300–305
122. Roomi MW, Gaal K, Yuan QX, French BA, Fu P, Bardag-Gorce F, French SW (2006) Preneoplastic liver cell foci expansion induced by thioacetamide toxicity in drug-primed mice. Exp Mol Pathol 81(1):8–14
123. Borud O, Mortensen B, Mikkelsen IM, Leroy P, Wellman M, Huseby NE (2000) Regulation of gamma-glutamyltransferase in cisplatin-resistant and -sensitive colon carcinoma cells after acute cisplatin and oxidative stress exposures. Int J Cancer 88(3):464–468
124. Pankiv S, Moller S, Bjorkoy G, Moens U, Huseby NE (2006) Radiation-induced upregulation of gamma-glutamyltransferase in colon carcinoma cells is mediated through the Ras signal transduction pathway. Biochim Biophys Acta 1760(2):151–157
125. Pandur S, Pankiv S, Johannessen M, Moens U, Huseby NE (2007) Gamma-glutamyltransferase is upregulated after oxidative stress through the Ras signal transduction pathway in rat colon carcinoma cells. Free Radical Res 41(12):1376–1384
126. Daubeuf S, Accaoui MJ, Pettersen I, Huseby NE, Visvikis A, Galteau MM (2001) Differential regulation of gamma-glutamyltransferase mRNAs in four human tumour cell lines. Biochim Biophys Acta 1568(1):67–73
127. Bouman L, Sanceau J, Rouillard D, Bauvois B (2002) Gamma-glutamyl transpeptidase expression in Ewing's sarcoma cells: up-regulation by interferons. Biochem J 364(Pt 3): 719–724
128. Reuter S, Schnekenburger M, Cristofanon S, Buck I, Teiten MH, Daubeuf S, Eifes S, Dicato M, Aggarwal BB, Visvikis A, Diederich M (2009) Tumor necrosis factor alpha induces gamma-glutamyltransferase expression via nuclear factor-kappaB in cooperation with Sp1. Biochem Pharmacol 77(3):397–411
129. Corti A, Franzini M, Paolicchi A, Pompella A (2010) Gamma-glutamyltransferase of cancer cells at the crossroads of tumor progression, drug resistance and drug targeting. Anticancer Res 30(4):1169–1181
130. Sze G, Kaplowitz N, Ookhtens M, Lu SC (1993) Bidirectional membrane transport of intact glutathione in Hep G2 cells. Am J Physiol 265(6 Pt 1):G1128–G1134
131. Hanigan MH (1995) Expression of gamma-glutamyl transpeptidase provides tumor cells with a selective growth advantage at physiologic concentrations of cyst(e)ine. Carcinogenesis 16(2):181–185
132. Hanigan MH, Gallagher BC, Townsend DM, Gabarra V (1999) Gamma-glutamyl transpeptidase accelerates tumor growth and increases the resistance of tumors to cisplatin in vivo. Carcinogenesis 20(4):553–559
133. Franzini M, Corti A, Lorenzini E, Paolicchi A, Pompella A, De Cesare M, Perego P, Gatti L, Leone R, Apostoli P, Zunino F (2006) Modulation of cell growth and cisplatin sensitivity by membrane gamma-glutamyltransferase in melanoma cells. Eur J Cancer 42(15):2623–2630
134. Paolicchi A, Lorenzini E, Perego P, Supino R, Zunino F, Comporti M, Pompella A (2002) Extra-cellular thiol metabolism in clones of human metastatic melanoma with different gamma-glutamyl transpeptidase expression: implications for cell response to platinum-based drugs. Int J Cancer 97(6):740–745

135. Kroning R, Lichtenstein AK, Nagami GT (2000) Sulfur-containing amino acids decrease cisplatin cytotoxicity and uptake in renal tubule epithelial cell lines. Cancer Chemother Pharmacol 45(1):43–49
136. Paolicchi A, Sotiropuolou M, Perego P, Daubeuf S, Visvikis A, Lorenzini E, Franzini M, Romiti N, Chieli E, Leone R, Apostoli P, Colangelo D, Zunino F, Pompella A (2003) Gamma-glutamyl transpeptidase catalyses the extracellular detoxification of cisplatin in a human cell line derived from the proximal convoluted tubule of the kidney. Eur J Cancer 39(7):996–1003
137. Dominici S, Valentini M, Maellaro E, Del Bello B, Paolicchi A, Lorenzini E, Tongiani R, Comporti M, Pompella A (1999) Redox modulation of cell surface protein thiols in U937 lymphoma cells: the role of gamma-glutamyl transpeptidase-dependent H_2O_2 production and S-thiolation. Free Radical Biol Med 27(5–6):623–635
138. Corti A, Duarte TL, Giommarelli C, De Tata V, Paolicchi A, Jones GD, Pompella A (2009) Membrane gamma-glutamyltransferaseactivitypromotesiron-dependentoxidative DNA damage in melanoma cells. Mutat Res 669(1–2):112–121
139. Dominici S, Pieri L, Comporti M, Pompella A (2003) Possible role of membrane gamma-glutamyltransferase activity in the facilitation of transferrin-dependent and -independent iron uptake by cancer cells. Cancer Cell Int 3(1):7
140. Maellaro E, Dominici S, Del Bello B, Valentini MA, Pieri L, Perego P, Supino R, Zunino F, Lorenzini E, Paolicchi A, Comporti M, Pompella A (2000) Membrane gamma-glutamyl transpeptidase activity of melanoma cells: effectsoncellular H_2O_2 production, cellsurface protein thiol oxidation and NF-kappa B activation status. J Cell Sci 113(Pt 15):2671–2678
141. Corti A, Paolicchi A, Franzini M, Dominici S, Casini AF, Pompella A (2005) The S-thiolating activity of membrane gamma-glutamyltransferase: formation of cysteinyl-glycine mixed disulfides with cellular proteins and in the cell microenvironment. Antioxid Redox Signal 7(7–8):911–918
142. Castellano I, Ruocco MR, Cecere F, Di Maro A, Chambery A, Michniewicz A, Parlato G, Masullo M, De Vendittis E (2008) Glutathionylation of the iron superoxide dismutase from the psychrophilic eubacterium *Pseudoalteromonas haloplanktis*. Biochim Biophys Acta 1784(5):816–826
143. Dominici S, Visvikis A, Pieri L, Paolicchi A, Valentini MA, Comporti M, Pompella A (2003) Redox modulation of NF-kappaB nuclear translocation and DNA binding in metastatic melanoma. The role of endogenous and gamma-glutamyl transferase-dependent oxidative stress. Tumori 89(4):426–433
144. Dominici S, Pieri L, Paolicchi A, De Tata V, Zunino F, Pompella A (2004) Endogenous oxidative stress induces distinct redox forms of tumor necrosis factor receptor-1 in melanoma cells. Ann N Y Acad Sci 1030:62–68
145. Pieri L, Dominici S, Del Bello B, Maellaro E, Comporti M, Paolicchi A, Pompella A (2003) Redox modulation of protein kinase/phosphatase balance in melanoma cells: the role of endogenous and gamma-glutamyltransferase-dependent H_2O_2 production. Biochim Biophys Acta 1621(1):76–83
146. Giommarelli C, Corti A, Supino R, Favini E, Paolicchi A, Pompella A, Zunino F (2008) Cellular response to oxidative stress and ascorbic acid in melanoma cells overexpressing gamma-glutamyltransferase. Eur J Cancer 44(5):750–759
147. Sullivan R, Graham CH (2008) Chemosensitization of cancer by nitric oxide. Curr Pharm Des 14(11):1113–1123
148. Frederiksen LJ, Sullivan R, Maxwell LR, Macdonald-Goodfellow SK, Adams MA, Bennett BM, Siemens DR, Graham CH (2007) Chemosensitization of cancer in vitro and in vivo by nitric oxide signaling. Clin Cancer Res 13(7):2199–2206
149. Weyerbrock A, Baumer B, Papazoglou A (2009) Growth inhibition and chemosensitization of exogenous nitric oxide released from NONOates in glioma cells in vitro. J Neurosurg 110(1):128–136

150. Adams C, McCarthy HO, Coulter JA, Worthington J, Murphy C, Robson T, Hirst DG (2009) Nitric oxide synthase gene therapy enhances the toxicity of cisplatin in cancer cells. J Gene Med 11(2):160–168

151. Bramanti E, Angeli V, Franzini M, Vecoli C, Baldassini R, Paolicchi A, Barsacchi R, Pompella A (2009) Exogenous vs. endogenous gamma-glutamyltransferaseactivity: implications for the specificdetermination of S-nitrosoglutathione in biologicalsamples. Arch Biochem Biophys 487(2):146–152

152. Lee DH, Silventoinen K, Jacobs DR Jr, Jousilahti P, Tuomileto J (2004) Gamma-glutamyltransferase, obesity, and the risk of type 2 diabetes: observational cohort study among 20,158 middle-aged men and women. J Clin Endocrinol Metab 89(11):5410–5414

153. Kazemi-Shirazi L, Endler G, Winkler S, Schickbauer T, Wagner O, Marsik C (2007) Gamma glutamyltransferase and long-term survival: is it just the liver? Clin Chem 53(5):940–946

154. Franzini M, Corti A, Fornaciari I, Balderi M, Torracca F, Lorenzini E, Baggiani A, Pompella A, Emdin M, Paolicchi A (2009) Cultured human cells release soluble gamma-glutamyltransferase complexes corresponding to the plasma b-GGT. Biomarkers 14(7):486–492

155. Li X, Mortensen B, Rushfeldt C, Huseby NE (1998) Serum gamma-glutamyltransferase and alkaline phosphatase during experimental liver metastases. Detection of tumour-specific isoforms and factors affecting their serum levels. Eur J Cancer 34(12):1935–1940

156. Pettersen I, Andersen JH, Bjornland K, Mathisen O, Bremnes R, Wellman M, Visvikis A, Huseby NE (2003) Heterogeneity in gamma-glutamyltransferase mRNA expression and glycan structures. Search for tumor-specific variants in human liver metastases and colon carcinoma cells. Biochim Biophys Acta 1648(1–2):210–218

157. Simic T, Dragicevic D, Savic-Radojevic A, Cimbaljevic S, Tulic C, Mimic-Oka J (2007) Serum gamma glutamyl-transferase is a sensitive but unspecific marker of metastatic renal cell carcinoma. Int J Urol 14(4):289–293

158. Strasak AM, Rapp K, Brant LJ, Hilbe W, Gregory M, Oberaigner W, Ruttmann E, Concin H, Diem G, Pfeiffer KP, Ulmer H (2008) Association of gamma-glutamyltransferase and risk of cancer incidence in men: a prospective study. Cancer Res 68(10):3970–3977

159. Mena S, Benlloch M, Ortega A, Carretero J, Obrador E, Asensi M, Petschen I, Brown BD, Estrela JM (2007) Bcl-2 and glutathione depletion sensitizes B16 melanoma to combination therapy and eliminates metastatic disease. Clin Cancer Res 13(9):2658–2666

160. Hiramatsu K, Asaba Y, Takeshita S, Nimura Y, Tatsumi S, Katagiri N, Niida S, Nakajima T, Tanaka S, Ito M, Karsenty G, Ikeda K (2007) Overexpression of gamma-glutamyltransferase in transgenic mice accelerates bone resorption and causes osteoporosis. Endocrinology 148(6):2708–2715

161. Suzuki H, Izuka S, Minami H, Miyakawa N, Ishihara S, Kumagai H (2003) Use of bacterial gamma-glutamyltranspeptidase for enzymatic synthesis of gamma-D-glutamyl compounds. Appl Environ Microbiol 69(11):6399–6404

162. Simbirtsey A, Kolobov A, Zabolotnych N, Pigareva N, Konusova V, Kotov A, Variouchina E, Bokovanov V, Vinogradova T, Vasilieva S, Tuthill C (2003) Biological activity of peptide SCV-07 against murine tuberculosis. Russ J Immunol 8:11–22

163. Rose WA 2nd, Tuthill C, Pyles RB (2008) An immunomodulating dipeptide, SCV-07, is a potential therapeutic for recurrent genital herpes simplex virus type 2 (HSV-2). Int J Antimicrob Agents 32(3):262–266

164. Suzuki H, Kato K, Kumagai H (2004) Development of an efficient enzymatic production of gamma-d-glutamyl-L-tryptophan (SCV-07), a prospective medicine for tuberculosis, with bacterial gamma-glutamyltranspeptidase. J Biotechnol 111:291–295

165. Jolivet S, Mooibroek H, Wichers HJ (1998) Space-time distribution of γ-glutamyl transferase activity in *Agaricus bisporus*. FEMS Microbiol Lett 163:263–267

166. Zhang H, Zhan Y, Chang J, Liu J, Xu L, Wang Z, Liu Q, Jiao Q (2012) Enzymatic synthesis of b-N-(c-L(+)-glutamyl)phenylhydrazine with *Escherichia coli* gamma-glutamyltranspeptidase. Biotechnol Lett 34:1931–1935

167. Zhang H (2013) Enzymatic synthesis of β-N-(γ-l(+)-glutamyl)-4-carboxyphenylhydrazine with *Escherichia coli* γ-glutamyltransferase. J Mol Catal B Enzym 90:128–131

168. Tischer W, Giesecke U, Lang G, Roder A, Wedekind F (1992) Biocatalytic 7-aminocephaiosporanic acid production. Ann N Y Acad Sci 672:502–509

169. Li Y, Chen J, Jiang W, Mao X, Zhao G, Wang E (1999) In vivo post-translational processing and subunit reconstitution of cephalosporin acylase from *Pseudomonas* sp. 130. Eur J Biochem 262(3):713–719

170. Suzuki H, Yamada C, Kijima K, Ishihara S, Wada K, Fukuyama K, Kumagai H (2010) Enhancement of glutaryl-7-aminocephalosporanic acid acylase activity of gamma-glutamyltranspeptidase of *Bacillus subtilis*. Biotechnol J 5(8):829–837

171. Vermeulen N, Gänzle MG, Vogel RF (2007) Glutamine deamidation by cereal-associated lactic acid bacteria. J Appl Microbiol 103:1197–1205

172. Sakato Y (1949) The chemical constituents of tea III. A new amide theanine. J Agric Food Chem 23:262–267

173. Jujena LR, Chu D, Okubo T, Nagato Y, Yokogoshi H (1999) L-theanine—a unique amino acid of green tea and its relaxation effect in humans. Trends Food Sci Technol 10(6–7):199–204

174. Yokogoshi H, Kato Y, Sagesaka YM, Takihara-Matsuura T, Kakuda T, Takeuchi N (1995) Reduction effect of theanine on blood pressure and brain 5-hydroxyindoles in spontaneously hypertensive rats. Biosci Biotechnol Biochem 59(4):615–618

175. Park SK, Jung IC, Lee WK et al (2011) A combination of green tea extract and l-theanine improves memory and attention in subjects with mild cognitive impairment: a double-blind placebo-controlled study. J Med Food 14(4):334–343

176. Giesbrecht T, Rycroft JA, Rowson MJ, De Bruin EA (2010) The combination of L-theanine and caffeine improves cognitive performance and increases subjective alertness. Nutr Neurosci 13(6):283–290

177. Sadzuka Y, Sugiyama T, Sonobe T (2000) Efficacies of tea components on doxorubicin induced antitumor activity and reversal of multidrug resistance. Toxicol Lett 114(1–3):155–162

178. Kimura K, Ozeki M, Juneja LR, Ohira H (2007) L-theanine reduces psychological and physiological stress responses. Biol Psychol 74(1):39–45

179. Egashira N, Hayakawa K, Mishima K, Kimura H, Iwasaki K, Fujiwara M (2004) Neuroprotective effect of gamma-glutamylethylamide (theanine) on cerebral infarction in mice. Neurosci Lett 363(1):58–61

180. Zhang ZZ, Yan SH, Li DX, Jing TI, Meurens M, Larondelle Y (2011) Chemical synthesis and the stability of theanine. Adv Mater Res 396–398:1273–1277

181. Matsuura T, Kakuda T (1990) Effect of precursor, temperature, and illumination on theanine accumulation in tea callus. Agric Biol Chem 54:2283–2286

182. Li J, Li P, Liu F (2008) Production of theanine by *Xerocomus badius* using submerged fermentation. LWT 41:883–889

183. Lichtenstein N (1942) Preparation of c-alkylamides of glutamic acid. J Am Chem Soc 64:1021–1022

184. Kawagishi H, Sugiyama K (1992) Facile and large-scale synthesis of L-theanine. Biosci Biotechnol Biochem 56:689

185. Yan SH, Dufour JP, Meurens M (2003) Synthesis and characterization of highly pure theanine. J Tea Sci 23:99–104

186. Suzuki H, Izuka S, Miyakawa N, Kumagai H (2002) Enzymatic production of theanine, an 'umami' component of tea, from glutamine and ethylamine with bacterial γ-glutamyltranspeptidase. Enzyme Microb Technol 31:884–889

187. Hogg N, Singh RJ, Konorev E, Joseph J, Kalyanaraman B (1997) S-Nitrosoglutathione as a substrate for gamma-glutamyl transpeptidase. Biochem J 323(Pt 2):477–481

188. Schrier RW, Wang W, Poole B, Mitra A (2004) Acute renal failure: definitions, diagnosis, pathogenesis, and therapy. J Clin Invest 114(1):5–14

189. Zhang Q, Kulczynska A, Webb DJ, Megson IL, Botting NP (2013) A new class of NO-donor pro-drugs triggered by gamma-glutamyl transpeptidase with potential for reno-selective vasodilatation. Chem Commun (Camb) 49(14):1389–1391
190. Wilk S, Mizoguchi H, Orlowski M (1978) Gamma-glutamyl dopa: a kidney-specific dopamine precursor. J Pharmacol Exp Ther 206(1):227–232
191. Di Stefano A, Sozio P, Cerasa LS (2008) Antiparkinson prodrugs. Molecules 13:46–68
192. Lu SC (1999) Regulation of hepatic glutathione synthesis: current concepts and controversies. FASEB J 13(10):1169–1183